AF745325

Sustainable Business
Integrating CSR in Business and Functions

RIVER PUBLISHERS SERIES IN MULTI BUSINESS MODEL INNOVATION, TECHNOLOGIES AND SUSTAINABLE BUSINESS

Series Editors

PETER LINDGREN
Aarhus University
Denmark

ANNABETH AAGAARD
Aarhus University
Denmark

The River Publishers Series in Multi Business Model Innovation, Technologies and Sustainable Business includes the theory and use of multi business model innovation, technologies and sustainability involving typologies, ontologies, innovation methods and tools for multi business models, and sustainable business and sustainable innovation. The series cover cross technology business modeling, cross functional business models, network based business modeling, Green Business Models, Social Business Models, Global Business Models, Multi Business Model Innovation, interdisciplinary business model innovation. Strategic Business Model Innovation, Business Model Innovation Leadership and Management, technologies and software for supporting multi business modeling, Multi business modeling and strategic multi business modeling in different physical, digital and virtual worlds and sensing business models. Furthermore the series includes sustainable business models, sustainable & social innovation, CSR & sustainability in businesses and social entrepreneurship.

Key topics of the book series include:

- Multi business models
- Network based business models
- Open and closed business models
- Multi Business Model eco systems
- Global Business Models
- Multi Business model Innovation Leadership and Management
- Multi Business Model innovation models, methods and tools
- Sensing Multi Business Models
- Sustainable business models
- Sustainability & CSR in businesses
- Sustainable & social innovation
- Social entrepreneurship and -intrapreneurship

For a list of other books in this series, visit www.riverpublishers.com

Sustainable Business
Integrating CSR in Business and Functions

Annabeth Aagaard

Department of Business Development and Technology
Aarhus University
Herning, Denmark

River Publishers

Published, sold and distributed by:
River Publishers
Alsbjergvej 10
9260 Gistrup
Denmark

River Publishers
Lange Geer 44
2611 PW Delft
The Netherlands

Tel.: +45369953197
www.riverpublishers.com

ISBN: 978-87-93379-79-4 (Hardback)
978-87-93379-80-0 (Ebook)

©2016 River Publishers

All rights reserved. No part of this publication may be reproduced, stored in a retrieval system, or transmitted in any form or by any means, mechanical, photocopying, recording or otherwise, without prior written permission of the publishers.

I dedicate this book to my children, Kristian and Viktoria.

Contents

Acknowledgments

Writing a book is a long journey, not a lonely trip, but one that you share with a number of brilliant people, who continuously inspire and motivate you along your venture.

First of all, I would like to thank my two kids, my husband, my sister and my parents for always supporting me in whatever project or idea I want to pursue. Without you, none of this would matter.

Second, I would like to thank all of the inspiring CSR representatives of some of the key sustainable businesses across the world, who have been so kind to share their experiences and advises on successful CSR and sustainable business integration. Without them, this book would have been just another theoretical textbook. However, with their help and specialist knowledge, this book will contribute in understanding 'how' sustainable businesses are created and integrated in theory and practice. The people are:

- Henkel: General Manager of Henkel Laundry & Home Care France, Jean-Baptiste Santoul and head of sustainability management, Uwe Bergman
- LEGO: CSR director Helle Kaspersen
- Novo Nordisk: Vice president of Corporate Sustainability, Susanne Stormer
- RSA and Codan: Nordic Director Corporate Communications & CSR, Helle Fangel Løgstrup
- Hummel and Thornico: CEO Christian Stadil
- Rynkeby Foods: Quality manager Carina Jensen and Quality/CSR manager Rikke Bekker Henriksen
- Aalborg Municipality: Head of Health and Sustainable Development, Thomas Kastrup-Larsen
- The Specialists and the Specialist People Foundation, COO Henrik Thomsen

Third, I would like to thank colleagues at Aarhus University, University of Southern Denmark and Stanford University for your inspiration and input for the different chapters, research, and theoretical models. Without them, I would not have been able to dig as deep into state-of-the-art research across the CSR and sustainability research field.

Last, but not least, I would like to thank Rajeev Prasad, Junko Nakajima, and River Publishers. Without them this would have been a very lonely journey. So thanks for helping me in creating this book that will hopefully inspire lots of professionals, organizations, academics, and students on sustainable business and successful CSR integration across functions and organizations.

List of Figures

List of Tables

List of Abbreviations

BM	Business Model
BMI	Business Model Innovation
CS	Corporate Sustainability
CSI	Corporate Social Innovation
CSR	Corporate Social Responsibility
CR	Corporate Responsibility
HR	Human Resources
HRM	Human Resource Management
NGO	Non-Governmental Organization
R&D	Research and Development
RPM	Responsible Procurement Management
SME	Small- and Medium-sized Entities
TQM	Total Quality Management

Introduction

The developments in our environment and society have revealed that the way we run our businesses and govern our nations is not sustainable in the long run. Never before has environmentalists spoken so negatively on the outlook of the environment. The footprint of corporations and nations' continuous resource overuse and pollution can be seen and measured in nature. Similarly, the global world crisis exposed a decade-long economic irresponsibility among several nations, where social developments clearly show that the rich got richer and the poorer got poorer. These developments are questioning the accuracy of and accountability at many of the fundamental ways we have learned and think about business, where monetary profit for many years has been the single most important parameter for assessing the success of the business, whereas the way in which the company has created its profit has been secondary or of even lower priority. Similarly, many nations and companies have for many years mainly acted on their own economic needs without any business interests for the environment, society, and/or their stakeholders. The consequences of this strongly emphasize the need for companies and governments to change their behavior quickly and efficiently toward more sustainable ways of doing business. Many global environmental and aid organizations, NGO's[1], and scientists have for years tried to create awareness and dialogue about these consequences, and more and more companies and nations are now beginning to act more responsibly rather than just talking about it. The global developments in the environment and the financial crisis made the interconnectedness between nations and their businesses very visible, underlining the need to think about liability in a more global and holistic fashion. Running a business today entails managing supply chains and stakeholders across nations with different cultures and living standards. Maneuvering across these political and societal systems is a balancing act. One thing is avoiding trouble; another approach is seeing the potentials herein and creating sustainable business and innovation. We are moving toward a new era of business, which is more connected and where collaborations across society,

[1]Non-governmental organizations (NGO).

public organizations, NGO's, and businesses can create new and sustainable business opportunities.

With the global media platforms such as Youtube, Facebook, and Twitter allowing for more transparency, companies cannot hide anymore. This again implies that customers can make more sustainable choices and put pressure on companies to produce in more sustainable ways and to introduce more sustainable products and services. At the same time, a new generation, Generation Y, is making its debut at the labor market, a generation that makes new demands on companies and their leaders than previous generations, in relation to social and environmental responsibility. Customers and the new generation of employees therefore require greater accountability and prioritize differently than previous generations of customers and employees, which stresses a different management approach in organizing businesses to be more sustainable. In businesses and organizations, this is reflected in new requirement and ways of involving internal and external stakeholders, a more holistic understanding of leadership and work/life balance as well as a more ethical, environment-friendly, and philanthropic approaches toward running a sustainable business and making money in a sustainable way.

The majority of the world's large, global organizations are already working with sustainability, and many have explicit strategies for their corporate social responsibility. However, the challenge for many organizations relates to integrating the corporate social responsibility (CSR) across the specific functions of the company to ensure the full business potential of sustainability. Many times, corporate sustainability remains corporate and is not well-integrated into sales, marketing, production, procurement, R&D, service, and administration. Without effective integration across the value chain of the organization, CSR will really just remain a marketing strategy and not generate the unique and new opportunities for sustainable business, and sustainable innovations and developments for the company.

Porter and Kramer (2006) emphasize the need for linking CSR, sustainable business, and competitive advantage: "Addressing social issues by creating shared value will lead to self-sustaining solutions that do not depend on private or government subsidies. When a well-run business applies its vast resources, expertise, and management talent to problems that it understands and in which it has a stake, it can have a greater impact on social good than any other institution or philanthropic organization." A key element in corporate sustainability, which has taken on a role of its own in research as well as practice, is philanthrophy. Many multinational companies are keen spokespersons for philanthrophy and apply is vividly in their corporate communication. However, businesses cannot survive over time, if they do not make money. Thus, sustainability and business have to be mixed well for companies to be able to survive, grow, and develop over time.

The Aim and Target Group

From research and practice cases, it appears that there is not one right way of working with and integrating sustainability into business and functions. The unique characteristics of the organizations, their sizes, their value chains, and the industrial and cultural context do have an impact on how CSR is integrated into business. This is also why this book contains cases from different types, sizes, and industrial contexts of companies and organizations, as a way to emphasize the various applications of sustainable business. It is therefore the aim of this book publication to provide knowledge of how to facilitate sustainability in business by integrating CSR into different types of business and functions.

More specifically, the book will answer the following: (1) What is CSR and sustainable business? (2) How can we integrate CSR into business? (3) How can we implement CSR and sustainable business into strategy? (4) In what way is CSR and sustainable business integrated into company functions (sales and communication, production and procurement, R&D, HRM, and administration)? (5) How can CSR and sustainable business be integrated and facilitated in different industries? And (6) What does the future hold for sustainable businesses? These questions will be answered throughout the different chapters of the book using theories, models, and practice cases.

The target group for the book is universities and other educational institutions as well as public and private organizations and managers that want to explore and succeed with the integration of CSR into their business, organization, and its functions.

How to Read and Apply this Book

It is important to emphasize that this book is not a complete theoretical compendium to CSR or sustainable business. "*Sustainable business—integrating CSR in business and functions*" is a research-based and applied-focused textbook of how to understand and integrate CSR and sustainability into business and company functions. Through the application of research and practice cases, the book attempts to bridge between theory and practice providing knowledge of how to facilitate sustainable business by integrating

CSR in business and across company functions. This is done with the main objective to ensure a better strategic fit between business and sustainability across organizations.

The book may be used as textbook on bachelor and master's level in courses emphasizing CSR, sustainability, sustainable business, and strategic management. As a suggestion, the cases may be applied in workshops, where the lecturer suggests a number of questions related to theories of CSR and sustainability, to be answered by the students during the course. As the book builds on numerous theoretical papers, specific papers may be selected from the list of references to be applied as supplementary reading for the students or as part of article workshops, where the students present the papers addressing key elements in the theoretical CSR discussion. Furthermore, the book can be applied as guide for organizations and businesses to implement CSR more effectively into their business and across the value chain and functions to ensure more optimal and sustainable business results.

1

The Concepts and History of CSR and Sustainability

In 1983, the UN decided to establish a World Commission on Environment and Development, headed by the former prime minister Gro Harlem Brundtland as chairman of the Board. In 1987, the Commission submitted its report to the general assembly of the United Nations with the title "Our Common Future." In the Brundtland report, as it was called, sustainable development is defined as follows:

Sustainable development is a development that meets the needs of the present without compromising the ability of future generations to meet their own needs.

The Brundtland report gave the kickoff to a worldwide interest for the environment, which subsequently to a large extent has set the agenda for both governments' and consumers' environmental considerations and thus also the industry's terms and conditions. In the business community, this has among other resulted in higher demands for environmental approvals for polluting enterprises and the introduction of a number of green taxes.

Globalization appears to be one of the key driving forces behind CSR. With its emergence, presently operating companies have managed to build stronger connections than before across international, environmental, social, and ethical boundaries and legislations, surpassing geographical barriers. Several large companies consist of a myriad of global collaborations, stakeholders, suppliers, and sub-suppliers where resources, employees, and raw materials come from various distinguished places in the world. This interconnectedness calls for greater demands on corporate sustainability across the companies' many stakeholders as the consequences of non-sustainable behavior affects everybody in the global supply chain. Simultaneously, the development in information technology, the global media, and the power of customers and NGOs emphasizes on sustainable business. There remains no scope for companies to hide and run from their responsibilities. By extension, the multinationals have emerged as even bigger and more powerful, which places them in an even stronger position in the national and global economy and in the global political debate. It is therefore necessary that these companies understand and assume the responsibilities that come with such an economic and political power.

1.1 The Brundtland Report

Sustainability is considered by the Brundtland report to be necessary for the entire world, both for the industrialized and non-industrialized countries. According to the Brundtland report, although the industrialized world amounts to only 20% of the world's population, it consumes 80% of the world's resources, an undeniable count. The development in the industrialized countries therefore significantly affects the non-industrialized countries, because they use as large a part of the world's resources. The report provides a comprehensive overview of the largest global environmental crises and provides suggestions for how to solve the problems. The major global challenges facing humanity, as presented by the report, constitute the following (WCED, 1987):

1. Population and human resources
2. Food security
3. Species and ecosystems
4. Energy
5. Industry
6. City environment

The correlation between the environment and development is a central part in the report, as these two factors are considered mutually dependent and are thus explained in the following manner: "The 'environment' is where we all live; and 'development' is what we all do in attempting to improve our lot within that abode" (WCED, 1987). The Commission has submitted a report of 22 new principles for legislation, which may help to create a sustainable development. The report recommends that these principles are incorporated into national laws, regulations, or court documents, which set out rights and obligations for citizens and States, and also that they are written into a global convention on States' sovereign rights and responsibilities. World politicians in the report are asked to use eight important international objectives as guidelines for their work (WCED, 1987):

1. Restoration of economic growth.
2. Improvement of the quality of growth to ensure the environmental and social common sense and that the needs for work, food, water, and sanitary improvements will be covered.

3. Conservation and enhancement of the natural resource base.
4. Stabilization of population figures.
5. Diversion of technology and improved risk management.
6. Integration of the environment and the economy in the decision-making process.
7. Improvement of global economic conditions.
8. Strengthening of international cooperation.

The Brundtland report has been fundamental to the establishment, development, and integration of the way we today understand sustainable business and CSR as a concept and as a way of doing business. According to professor Andrew Friedman and professor Samantha Miles (2002) and other researchers, organizations have proven to be a key factor in the process of creating a better world, and as a result, businesses have come, according to several scholars, under increased pressure to demonstrate good and responsible management (Pinkston and Carroll, 1994; De Bakker et al., 2005; Angus-Leppan et al., 2010). In most definitions, CSR emphasizes a company's responsibility to their stakeholders (Crook, 2005). This is represented by the triple bottom line "people, planet, profit" (Cramer et al., 2006), which brings together an extended range of values and criteria for the measurement of organizational success in the three groups: Economic, ecological, and social. The CSR definitions and the triple bottom line emphasize a more external focus for the company's actions, as the companies are not only set in the world to make a profit but also to maintain better social and environmental conditions. In practice, this means that the success of business was previously mainly measured using economic parameters, including financial reporting, which has been prepared on the basis of the requirements of the triple bottom line, addressing the company's environmental and social results. In other words, companies and businesses are no longer solely perceived as economic entities, but as social and ecological entities as well, which influence and are influenced by their surroundings. This consideration shines through companies' financial statements when they have been prepared in accordance with the principles behind the triple bottom line.

1.2 Defining the Concept of CSR and Corporate Sustainability

The two concepts, CSR and sustainability, are often applied interchangeably, and although there are extensions of each other, they are defined differently. A company's sustainability, also referred to as corporate sustainability, is considered as the foundation of CSR (Marrewijk, 2003). Corporate sustainability is defined by The World Business Council for Sustainable Development (WBCSD, 2000) as follows:

The business community's continued commitment to behave ethically and contribute to economic development and at the same time improving the quality of life for the employees and their families as well as for the local community and society as a whole.

The correlation between the two concepts is illustrated in Figure 1.1 by Marcel van Marrewijk, who is the director of the Dutch institute, "A great place to work," and who is one of the key contributors to CSR theory.

The concept of a company's CSR is defined as follows:

Corporate social responsibility includes the economic, legal, ethical and philanthropic expectations that society has to organizations at a given moment in time (Carroll, 1979: 500; Carroll and Shabana, 2010: 89).

This definition includes the time aspect, namely addressing the issue related to the fact that society's expectations of the organization gradually

Figure 1.1 The correlation between corporate sustainability and CSR.

Source: Marrewijk (2003).

change over time. This requires that companies do not just consider CSR as a static concept, but a fluid concept that they have to adjust their CSR according to making sure that their CSR activities meet the current needs and exceptions of society. Furthermore, corporate sustainability focuses more on the company and its surrounding and their impact on each other, whereas CSR focuses more on the stakeholders and the charitable and beneficial social activities that the company performs.

In a green paper from the EU Commission (2001), another definition of CSR is presented:

A company's voluntary work to integrate social and environmental concerns in their business operations and in interaction with their stakeholders.

This definition chiefly focuses on the voluntary work of integrating the social and environmental concerns into business and the interaction with the stakeholders. However, the companies' voluntary work with CSR nowadays is heavily affected by the instructions and requirements of CSR reporting, which are put forward by the UN and the EU to larger businesses. From the aforementioned and other existing definitions, it appears that CSR and (corporate) sustainability are not identical concepts, but rather complementary to each other. Archie B. Carroll, a retired professor from Terry College of Business, is one of the major proponents within the research area of CSR. His renowned CSR pyramid (Figure 1.2) draws up the CSR requirements of a company.

The CSR pyramid and Carroll's article from 1979 to 1991 show how certain requirements of a company is rooted in practice while other claims are desirable and expected today, but may potentially become statutory

Figure 1.2 The CSR pyramid.

Source: Carroll (1979, 1991).

requirements for the company in the near future. That a company can generate profit complying with national laws has always been a requirement, which the society searches for, and which the companies have to follow. That a company acts ethically correct is expected of society, and statutory on a few of the areas, but certainly not all. And that a company is responsible is desirable, but certainly not a statutory requirement. However, these things are in the process of constant change, surpassing the boundaries for sustainable businesses. More and more countries and their accounting laws are emphasizing and including requirements in relation to the sustainability of the companies. For example, Denmark in 2009 has introduced a new Financial Statements Act that indicates that all listed financial companies must report on CSR in their financial statements. In this way, the expected requests of society to sustainable behavior have now been transformed into legal requirements, which the companies must live up to. The growing requirements of society are therefore a force to be reckoned with as they will eventually become part of the laws and legislation over time.

1.3 CSR and Global Compact

UN Global Compact is the world's largest initiative for corporate social responsibility and has started in 1999 with the intention to involve private companies in solving some of the major social and environmental challenges of globalization in the world. Global Compact offers companies the opportunity to exchange their learning and experience with other companies. More than 8,000+ companies and 4,000 nonprofit organizations today[1] has joined the Global Compact network globally. The only thing required of the participating companies is that they annually report how they are implementing the Global Compact ten principles in their work. The ten principles are related to human rights, labor, environment, and anti-corruption[2]. Global Compact has evolved to be an international standard for sustainable business and for the world's businesses to operate in a decent manner. Global Compact is today not only the most widely used international initiative to promote private corporate social responsibility but also forms the basis for the work of implementing social responsibility in nonprofit organizations, public enterprises, and authorities. The ten principles include the key elements that the company should discuss

[1]www.unglobalcompact.org — downloaded May 14, 2016.

[2]See the ten principles in Appendix 2.

and attune to supplier relationships and collaborations, for example, integrated in the company's code of conduct. Code of conduct is the company's one of the guidelines for good practice on social, ethical, and environmental issues in terms of both the company's employees and the organization, as well as in relation to its suppliers[3].

In continuation of the UN Global Compact, the Global Reporting Initiative (GRI) has been established, which is a recognized initiative for reporting on corporate responsibility and sustainability. Global Compact and GRI cooperate to ensure consistency between principles and indicators of corporate social responsibility. The Global Compact consists of internationally agreed rules and conventions where GRI is a structure to account for economic, environmental, and social issues associated with the core indicators and additional indicators. The cornerstone of the GRI is a framework tool outlining principles and indicators that companies and organizations can use to measure and report their economic, environmental, and social performance.[4]

Several countries, including Britain, Sweden, Denmark, and France, have established legal requirements, ensuring that the largest companies report on CSR. These laws are targeting larger companies. However, these laws also have an impact on the small- and medium-sized companies. Partly because CSR initiatives have branched out into industries and organizations, and partly because the big companies, as part of their CSR reporting, now also set higher standards of sustainability to their suppliers, the growing number of companies who sign up for UN Global Reporting Initiative indicate that companies are increasingly assuming their responsibility and pursuing sustainable business.

1.4 Corporate Governance and Compliance

CSR is also closely associated with the concept of corporate governance. Corporate governance represents the company's choice of ethical standards, which are part of the company's compliance. Compliance is a common concept of the laws, regulations, norms, standards, and code of ethics, which companies choose to comply. Compliance is carried out not only to ensure that there are no violations but also to counter the risk that the company's business or reputation will be adversely affected (Cannon, 2012). Ensuring the compliance

[3]www.unglobalcompact.org

[4]https://www.globalreporting.org/services/preparation/Pages/default.aspx

of processes to legal regulations, governance guidelines, and strategic business requirements is a necessity in controlling business behavior. Compliance management is thus defined as follows:

Compliance management is the term referring to the definition of means to avoid such illegal actions by controlling an enterprise's activities. By extension, compliance management also refers to standards, frameworks, and software used to ensure the company's observance of legal texts. (Kharbili et al., 2008, p. 2).

Corporate governance is generally not regulated, but consists of a number of recommendations and "code of conducts." Originally, the focus of corporate governance was on the relationship between the management and the board of listed companies, whereas corporate governance today has a much broader perspective and is more about securing a broad and forward-looking strategic vision for the company (OECD, 2013). The OECD Principles of Corporate Governance were originally endorsed by OECD Ministers in 1999 and have since become an international benchmark for policymakers, investors, corporations, and other stakeholders worldwide. The principles have advanced the corporate governance agenda and provided specific guidance for legislative and regulatory initiatives in both OECD and non-OECD countries. The Financial Stability Forum has designated the principles as one of the 12 key standards for sound financial systems. The principles also provide the basis for an extensive program of cooperation between OECD and non-OECD countries. The OECD (2004) defines corporate governance in the following manner:

Corporate governance involves a set of relationships between a company's management, its board, its shareholders and other stakeholders. Corporate governance also provides the structure through which the objectives of the company are set, and the means of attaining those objectives and monitoring performance are determined.

The concepts of corporate governance and compliance are key elements in CSR and in managing a sustainable business. Several tools and guidelines for integrating and ensuring sustainable business are available for all organizations through the websites and publications by the UN Global Compact, Global Reporting Initiative (GRI), OECD, and the World Bank. The extensive level of national and international, public, and free materials about how to integrate sustainability into business makes it both possible and more accessible for companies to pursue CSR in their business and functions.

1.5 The Development in Management Theories—from Efficiency to Sustainability

Many of the management theories, which affect the way we today define and handle management, have their origin in the nineteenth century, where the American tradition of streamlining and efficiency management was found. This historic period was characterized by a growing industrialization, where the companies grew and became more powerful, the chief reason why the American anti-trust legislation at this time was initiated to reduce this power.

Frederick Winslow Taylor (1856–1915) is one of the founders of efficiency management movement and is considered to the founding father of the concept of scientific management, also called Taylorism. Taylorism represents a leadership theory, which analyzes effective working procedures. In 1898, Taylor started working at Bethlehem Steel where he and his team of assistants developed "high speed steel," which led to Taylor being awarded a gold medal at the world exhibition in Paris in 1901 and the Elliott Cresson medal at the Franklin Institute in Philadelphia. On the basis of his work, he published "The Principles of Scientific Management." Scientific management is based on Taylor's idea of "one best way" or one right way to do things, which according to Taylor can be mapped through scientific studies[5]. Taylor's scientific management is actually a "system for efficiency," and it is operating on the basis of the following four principles:

1. Replace rule of thumb with methods based on a scientific study of the tasks.
2. Select, train, and develop each employee rather than passively letting them train themselves.
3. Give each employee detailed instructions and supervision in how the employee should carry out his/her work.
4. Divide the work almost equally between managers and workers so that the managers can apply the principles behind scientific management to plan the work and so that the workers can carry out their tasks.[6]

Taylor's system of efficiency received criticism from many sides, among other from the union, and it accommodated several strikes where the system was implemented. However, from an objective viewpoint, scientific management, and with respect for what it has brought management theory, the system has shown, among other, how efficiency can be optimized through the structuring

[5]Frederick Taylor and Scientific Management on www.businessmate.org

[6]Frederick Taylor & Scientific Management on www.businessmate.org

of working methods and measurement methods. Also, many of the basic ideas behind taylorism can be traced in the efficiency and quality tools we use today, for example, lean and total quality management (TQM). Another central management theorist, who has affected our perception of the management concept, is Henri Fayol (1841–1925). His practical background as a Mining and Geological Engineer and later as a Director laid the foundation for the development of his general theory of business administration, which emerged 5 years after the establishment of the theory, but independent of Taylor's scientific management theory. Fayol's business administration theory is also known as fayolism and is the first comprehensive explanation of a general management theory. Fayol stressed five primary management functions, which had to be carried out as part of the effective management:

1. Planning
2. Organizing
3. Commands
4. Coordination
5. Control.

If you look at these functions, they constitute a significant part of the concepts and understanding of the management prevailing in both theory and practice today. There are also several common features in the management view that fayolism and taylorism present, which is why they are also often mentioned in continuation of each other. In continuation of the five primary management functions, Fayol also presented 14 management principles.[7]

A counterpoint to Taylorism and Fayolism came from Chester Barnard, whose massive impact on management and organizational theory is well-documented by several authors (e.g., Scott, 1987; Williamson, 1995; Mahoney, 2002). The thoughts that Chester Barnard presented in his work from 1938, "The Functions of the Executive," have been generated later in several schools within theoretical management and organizational research, for example, institutional theory, decision-making theory, and human relations theory (Mahoney, 2002). His influence can also be seen in HRM, communication, and the preparation of reward systems. The "Functions of the Executive" is based on Barnard's experience and centers around the manager's functions, including management communication and the interaction between authority and reward. In the work, he sums up the leader's most important functions in the following three actions:

[7] See appendix for Fayol's Fourteen Principles of Management.

1. Development and maintenance of a communication system
2. Protection of essential services from other members
3. Formulation of organizational goals and objectives[8]

Chester Barnard emphasizes that the manager's role is to be a professional, a role model for the company. In order for management to be effective, it must be regarded as legitimate. Barnard perceived that the central challenges for management were to balance the technological and the human dimensions of the organization. The managerial challenges consisted in communicating organizational targets and to win the cooperation of both the formal and informal organization. Accountability in the form of the honor and faithfulness with which the manager carries out his/her responsibilities was, according to Chester Barnard, the most important role for management (Gabor and Mahoney, 2010).

Barnard is particularly recognized for having introduced the systemic approach toward organizational studies where he stresses that "*for an organization to survive in the external environment and to succeed in the long run, it is necessary to maintain a co-operation with the employees to meet the conditions of efficiency* (Barnard, 1938, 56)".

Barnard defines that efficiency equal to "satisfaction of individual motives," and he emphasizes that the cooperative systems (companies) must create a surplus of satisfaction if they are to be effective. Barnard not only recognizes like Taylor the need for efficiency, but also focuses on the fact that *"while efficiency, which is the ultimate goal for common actions, is important, it is just as important to also satisfy individual motives (Barnard, 1938, 58)."* In other words, it is important that managers consider whether it is the right things, the company does, and not only on whether they do things quickly and efficiently.

Barnards recipe for success and survival for companies therefore consists of two related and mutually dependent processes: (1) those relating to the cooperative system (the company) as a whole and in relation to the surroundings and (2) those relating to the creation and distribution of satisfaction between individuals. He therefore ascribes instability and errors in the cooperation as defects in and/or between these two processes (Barnard, 1938: 60–61). Looking with today's eyes on these definitions and explanations, it appears that Chester Barnard is actually addressing what we today call sustainable management, value-based management, and welfare at work (Gehani, 2002).

[8]Mahoney, 2002.

The dominating business model and management approach of today, which is inspired by Taylor, Fayol, and other efficiency and management theorists, do overlook some of the key elements to sustainable business as stressed by Barnard. The focus of the companies on efficiency management has centered on quantity and not quality, and primarily with a short-term and not at long-term horizon for evaluation. The insufficient interest in internal and external sustainability and the long-term effects of this efficiency management approach are visible in many different areas. Not only has the consequences of a decade of excessive use of the world's natural resources revealed itself in nature but the welfare of workers is also diminishing, revealing that more and more employees are diagnosed with stress-related diseases and depression. According to the WHO, a minimum of 350 million people worldwide suffer from depression.[9] The report "Stress in America—Paying With Our Health" from 2015 is based on a survey conducted online in 2014 by Harris Poll on behalf of the American Psychological Association (APA), among 3,068 adults ages 18+ who reside in the United States. The results were weighted as needed for age, gender, race/ethnicity, education, region, and household income and revealed a number of alarming findings:

- 43% of the informants stated that stress has increased from the previous year.
- 46% reported of feeling depressed/sad due to stress in the previous month.
- The average stress level was at 4.9, and the stress level of women was increasing.

In a report from the Danish Health Department from 2007, it is concluded that mental workload is the cause of losses in the Danes' average lifespan of 6–7 months, as well as a significant reduction in the quality-adjusted life years (Juel et al., 2006), 30,000 hospitalizations, half a million contacts to general practitioners GPs, one million sick days due to work-related illness, almost 3,000 early retirements, and approximately $150 million in additional expenditures in the healthcare system as a result of the psychological workload (Juel et al., 2006).

All of these internal and external imbalances of the companies mainly focus on the need for at more holistic and sustainable management approaches and stakeholder management, which CSR and sustainable business are the answer to.

---

[9]Source: WHO—http://www.who.int/mediacentre/factsheets/fs369/en/ downloaded May 16, 2016.

1.5.1 Sustainable Business and Blue Ocean Strategy

The "old" worldview and management and business understanding, which in a century have affected companies, have spawned many imbalances in the environment and across the world that clearly demonstrate that new understandings and business models are needed. W. Chan Kim and Renee Mauborgne's concept of "blue ocean strategy" question the companies' current way of thinking and acting. The concept and book publication of the same name are a result of a comprehensive study of 150 strategic moves over the past centuries[10] within 30 different industries. The purpose of a blue ocean strategy is not to outperform competitors in the existing market, but to create a new market area or a "blue ocean" in order to make the competition irrelevant. The traditional approach to the market, by contrast, acted on competition in the same areas, which the book refers to as "red ocean strategy," as all competitors bleed under this strategy. The ideas behind the blue ocean strategy emphasize a sustainable mindset in relation to how we can see the world economy and businesses from a whole new perspective, focusing on how the elements of these ecosystems interact. Does it for example make any sense that Europeans are paying high prices for sugar and cereals through the European and protectionist customs duties when the third world can produce the same goods at a fraction of the price? From a protectionist point of view, it might make sense, as a key agenda of the EU is to be self-sufficient, but not if you consider sustainability and blue ocean strategy in relation to the global world economy and optimal use of world resources. So, how can the blue ocean strategy generate sustainable value for businesses?

An important element of blue ocean strategy is value innovation, which characterizes new ways to create value for customers, and how to tie up innovation to the company's values. If we start from the growing interest in sustainable products, the company's values of sustainability can support new and sustainable innovation that can assist in opening up new markets. Corporate social innovation (CSI) and sustainable innovation, as discussed later in the book, are based on the ability of companies to think sustainability into their innovation and the products and services that the company creates. It therefore makes perfect sense for a sustainable business to think green innovation when companies want to expand their markets and develop new product benefits (Kim and Mauborgne, 2005).

---

[10]From year 1880 to 2000.

1.5.2 Sustainable Business Requires Change Leadership

CSR requires that a business better understands how it affects and is affected by its surroundings and can act and change accordingly. Furthermore, as the society and the world economy changes continuously, a company's ability to change becomes a key competence in today's sustainable businesses. Thus, effective change management and leadership are prerequisites for successful CSR and sustainable business, but what makes change effectively? John Kotter describes the two concepts as follows:

Change management refers to the basic tools used to structure and control a change, whereas change leadership deals with the driving forces, visions and processes that support a major transformation (Kotter, 1990).

It is especially the manager's ability to build a permanent "change competence" among employees to ensure that current and future challenges and changes in and outside the company are used in better ways in developing new business opportunities (Hildebrandt and Brandi, 2005). As the renowned economist Paul Romer has stated: *A crisis is a terrible thing to waste.*[11] Even in times of crisis there are opportunities, and the ability to change and adapt to these options is a necessity. The companies that are able to adapt continuously to developments in technology and markets, the environment and society, and the customers and their ambient requirements and needs will always have a future.

Change management is the tool that has been applied by management in handling the "wave" of continuous and global changes that influence organizations and businesses. At this point, many companies have multiple change projects running at the same time, and seldom see the end of one change project before another change project has started. Yet the question is whether change management as a tool is equipped to handle the waves of change that affect global companies of today? Changes are in their nature seldom something a company can predict or manage. The idea of change management is originally inspired by Taylor's "one best way," but changes often require new skills to be handled or to harvest the business potential of the change and there is not just one best way to do that.

Management literature has for long emphasized the concept of *change readiness*; however, people may be ready for change, but not have the capacity to handle or maneuver in a changeable environment. In response to this, professor Anthony Buono from Bentley University has introduced a new

[11] Paul Romer was first quoted for this statement at a venture–capitalist meeting, November 2004 in California.

concept to the existing change management theory, namely *change capacity,* which is defined as follows:

The ability to implement and execute changes continuously in response to and acceptance of external changes (Buono et al., 2009).

Change capacity is a competence that has to be facilitated and trained as it is a necessary skill in today's changeable society and business world (Buono et al., 2009). Change capacity is also a prerequisite for successful, sustainable business as the changing needs of society and stakeholders and the influences from the global economy affect the opportunities and demands for sustainable business. It is therefore imperative that companies build this flexibility into their organizational design and in their way of managing their organizations if they want to be able to harvest the global, business potentials of sustainability (Aagaard, 2012).

2

Understanding Sustainable Business

In the rise of globalization, sustainability has become a new premise for implementing business—with a rapid shift from political discourses (Dryzek, 2005) into company boardrooms and company strategy (Bisgaard, 2009). However, studying the concept of sustainability is challenged by the fact that it is a fragmented concept where some researchers even question whether sustainability is a concept or a political discourse (Dryzek, 2005) or an artifact as suggested by Faber et al. (2005). Sustainability is referred to and applied in various ways in literature with terminologies such as corporate sustainability (CS), corporate responsibility (CR), corporate social responsibility (CSR), corporate governance, and sustainable development—just to mention the most frequently applied. In elaboration of the concepts, CSR is tightly interrelated with literature on business ethics, as explained in Carroll's (1999) historic literature review, emphasizing four levels of corporate responsibility: economic, legal, ethical, and philanthropic obligations in relation to being a "good" corporate citizen. The root of corporate sustainability exists at the social level as a political concept initiated through international political institutions and movements with the Brundtland report as the iconic founding stone (Hansen et al., 2009), whereas corporate responsibility (CR) is defined as activities beyond legal and economic obligations (Blowfield and Frynas, 2005). Corporate governance captures the management of different stakeholder demands and involves new types of activities such as voluntary code of conducts (Kourula and Halme, 2008).

2.1 The Concept of Sustainability and Sustainable Business

In spite of an existing growing body of literature in analyzing and discussing sustainability and sustainable development at the political and social level (Dryzek, 2005), the operationalization of the concept in relation to business is rather still weak (Bansal, 2005; Stubbs and Cocklin, 2008; Zink et al., 2008; Carroll and Shabana, 2010). What really matters is how to understand and operationalize sustainable business (SB) into corporate practices and performances creating and testing new sustainable opportunities and business models on corporate level (Holling, 2001; Newman, 2005). An attempt to transfer the general but rather vague Brundtland definition of sustainable development into corporate level is seen in Dyllick and Hockert's (2002: p. 131) definition of *sustainable development*:

Meeting the needs of a firm's direct and indirect stakeholders without compromising its ability to meet the needs of future stakeholders as well.

This explicit focus on stakeholder needs emphasizes the importance for companies not only to respond to their ecosystem and primary stakeholders such as shareholders, employees, and customers but also to secondary stakeholders such as NGOs in order to gain and maintain legitimacy and license-to-operate with regard to various sustainable issues (Zink et al., 2008). Furthermore, the application of a long-term perspective to various needs of future stakeholders underlines the complexity of long-term management practice and the short-term requests from shareholders on increased profits, which is a key challenge to be addressed at the corporate level (Poncelet, 2001).

However, the most common form of translation of sustainability on corporate level is the triple bottom line, which consists of three sustainable dimensions: people, planet, and profit (Elkington, 1997) and that is described as three equally important managerial principals (Hansen et al., 2009; Bradbury-Huang, 2010; Schaltegger and Wagner, 2011):

- The social dimension refers to equity for all human beings and their opportunities in gaining access to resources with regard to basic needs such as water, food, and development through improved living conditions such as health care and education (Bansal, 2005).
- The environmental dimension refers to the ecosystem of the Earth and to reductions of human created footprints and ecological imbalances in terms of pollution, the ozone layer, greenhouse gases, non-biodegradable waste, deforestation, and over fishing.
- The profit dimension emphasizes that production of goods and services are a prerequisite to improve the living conditions globally (Bansal, 2005).

The pressure on corporations to demonstrate their responsibility toward society has intensified over the past years, and the notion of CSR has institutionalized the ideal of an ethically alert organization able to balance its financial interests with its concern for society as a whole (Christensen et al., 2011). Although academic research has discussed the concept for many decades (Carroll, 1999), the field has primarily been driven by practitioners within large American corporations. The academic research in the field took off already in the 1950s (Carroll, 1999); still, however, there is enough confusion about how to define CSR and how it should be implemented (Welford, 2004). The field of CSR and sustainable business encompasses many different interests, goals, and voices and therefore is difficult to pin down precisely. The CSR literature flourishes with frameworks and approaches for identification and categorization of CSR systems. Some researchers (e.g., Melé, 2002; Philips et al., 2003) suggest that successful CSR has rather universalistic characteristics. Other researchers, such as Mitchell et al. (1997) and Swanson (1995), claim that the internal integration of CSR systems matters for their design. Yet, others (e.g., Husted, 2000) suggest that CSR systems should be contingent on contextual factors. These significantly different approaches to CSR have resulted in various foci, conceptual understandings, and definitions and have led to rather elaborate methodological and thematic relativism within the field.

One of most long-standing and widely cited definitions of CSR that has formed the basis of the more recent definitions comes from Carroll's (1979) "pyramid of CSR." Carroll offered the following definition:

> *The responsibility of business encompasses the economic, legal, ethical, and discretionary expectations that society has of organisations at a given point in time* (Carroll, 1979: 500).

The strength in this definition is its facilitation of the link between the organization and society through its combination of economic and discretionary expectations. Often, this is understood in a disintegrated manner with economic expectations deriving for the purpose of the organization and the discretionary expectations deriving for the purpose of others. While this distinction is attractive, Carroll's argument is integrative in the way that economic viability is performed also for the purpose of society. The economic responsibility is simply forming the base of the pyramid (Carroll, 1991). In this way, Carroll (1991) is both arguing for CSR as a legal and economic obligation to society and as an ethical obligation because CSR represents "the right thing to do."

Later definitions stress the integrated understanding of CSR, including all stakeholders and constituent groups of the corporation (Werther and

Chandler, 2005). For example, the EU Commission (2006) defines CSR as follows:

A concept whereby companies decide voluntarily to contribute to a better society and a cleaner environment. A concept whereby companies integrate social and environment concerns in their business operations and in their interaction with their stakeholders on a voluntary basis.

As such, the European Commission hedges its bets with two definitions wrapped into one and stresses the importance of integrating both social and environmental concerns in a firm's business operations. The definition also stresses the fact that a CSR initiative needs to be conducted on voluntary basis, which is beyond what is legally required. Business for social responsibility is broadly agreeing that CSR should focus on the impact of how core business is managed. Business for social responsibility (2000) defines CSR and sustainable business as follows:

Operating a business in a manner that meets or exceeds the ethical, legal, commercial and public expectations that society has of business.

The different definitions on CSR are mainly different in prescribing how far companies should go beyond managing their own impact into the terrain of acting specifically outside of that focus to make a contribution to the achievement of broader societal goals. The consequence of the different definitions is a morass of contestable models that makes CSR multiple disciplinary and composite with an array of theories used, which includes business studies, economics, sociology, politics, law, and philosophy. It is interesting to observe that none of the definitions actually defines sustainable business or the social responsibility of business, as so famously discussed by Milton Friedman (1970), but rather describe CSR as a phenomenon. This might be the cause of the definitional confusion: It is not so much a confusion of how CSR is defined, as it is about what constitutes the social responsibility of business. However, according to van Marrewijk (2003), a successful CSR strategy has to be context specific for each individual business, that is, what are the specific CSR issues to be addressed and how to engage with the stakeholders. Yet, a definition addressing these questions would not be applicable across a variety of contexts and thus would be less useful as a definition (Dahlsrud, 2008).

Corporate social responsibility (CSR) is an acknowledged and well-established strategic approach in theory and practice (Carroll and Buchholtz, 2006; Blowfield and Murray, 2008). Although several organizations have implemented CSR strategies, the level of organizational integration and the effects obtained through CSR varies (Porter and Kramer, 2006). Critiques of CSR have stressed the overemphasis on the externally oriented and often PR-driven approach to CSR in contrast to the internal integrated and employee-driven CSR approaches (Vogel, 2005; Kuhn and Deetz, 2008). Also, critiques have emphasized the lack of measured effects gained through CSR implementation (Friedman, 1970; Wright and Ferris, 1997; McWilliams and Siegel, 2000).

Corporate social responsibility has grown as a concept especially the last decade, theorists and practitioners have taken CSR to heart and developed it into an independent management discipline. The global need and widespread interest in sustainable development and sustainability of companies have made this a topic that all organizations have to deal with. The increased focus on corporate social responsibility is inevitably related to the structural changes in the global economy, helping to emphasize that the conditions for doing business is changing. Previously, corporate social responsibility largely emphasized philanthropy, whereas today it is a way of doing business. For the companies that have chosen to work with and integrate CSR, accountability becomes a critical element in corporate strategies, management, objectives, value chain, and collaborations. When companies pursue sustainable business, they must necessarily adjust and develop the way they do business and the way they make choices. This means that the company's stakeholders are given a much more central role in the definition and development of corporate strategy, core business, and product services. The primary objective of the company is suddenly no longer to make money in the short run, but to achieve sustainable success also in the long run and on an economic, social, and philanthropic level through the satisfaction and involvement of internal and external stakeholders and the community in the sustainable business of the company.

2.2 The Global Need for CSR and Sustainable Business

CSR is not a fad that will pass. For as long as there are inequalities and imbalances in our society and in the environment, there will be need for and commitment to sustainable business and CSR. This also means that CSR is becoming a general requirement of the public community to the way companies are doing business today and is in any event something that companies should consider. Bob McDonald, chairman and CEO of the American conglomerat Procter & Gamble, puts it this way:

I do not think that sustainability is an option anymore. The world today is so flat and transparent because of the Internet, and the influence of individuals' increased so much as a result of our ability to blogging and tweet and other things, so customers will want to know what they are buying into, when buying your brand. They want to know the company behind it. They want to know what the company stands for and how the company takes care of the environment" (quote from PWC, 2011: 13).

This statement stresses that CSR has become a business premise that companies find difficult to get around. The power that the customers have achieved through increased access to and dissemination of knowledge on the Internet has put even greater emphasis on the need to reveal "the true color" of a company—in relation to what the company stands for and what responsibilities it takes on. In the global CEO survey for 2011 by PWC, 1,200 corporate and government officials from 69 countries were asked whether they actively support policies that promote growth that is financially, socially, and environmentally sustainable. In response, 72% said yes. These data reveal a dominant trend that more and more business leaders actively support sustainable practices and prioritize to growth that is not only financial but also social and environmental.

The global development of CSR integration is often measured by the number of registered CSR reports, and the scope of CSR reporting across global companies has grown steadily since 1992 and appears to be unaffected by the global recession. CorporateRegister.com, which is the global registry for corporate responsibility (CR) reporting, could in June 2015 show that 19,192 in total had enrolled CSR reports (GRI, 2015).

Europe has had the lead on CSR reporting over the past 20 years and continues to produce a large portion of the registered CSR reports. Until 2009, UK was the most productive country in relation to CSR reporting, but in 2009 United States took the lead with the most reports per year. The development from 2007 to 2011 reveals that the Asian countries (particularly Japan and China) are gaining territory and that Brazil and Argentina are taking on a central role of sustainability in South America (Table 2.1).

2.3 Sustainable Business and Sustainable Business Model Innovation

Shifting from non-sustainable to sustainable business often requires alterations of existing business models and/or business model innovation. However, applying business model innovation as a way to create sustainable value requires several alterations of our ways of understanding and evaluating businesses and their business models. Both business model (BM) and business model innovation (BMI) have been the focus of substantial attention by

Table 2.1 The global division and development of CSR reporting

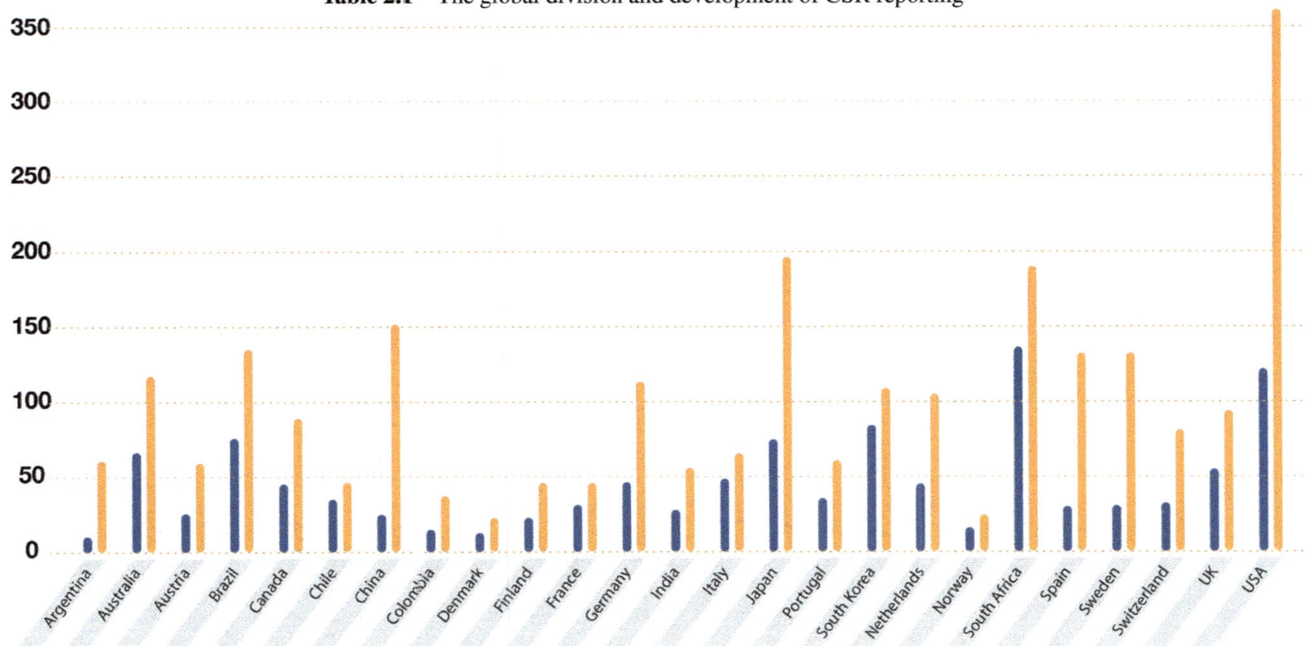

Source: KPMG (2013).

both academics and practitioners (e.g., Amit and Zott, 2001; Chesbrough and Rosenbloom, 2002; Christensen and Raynor, 2003; Govindarajan and Trimble, 2005; Markides, 2008; Teece, 2010; Lindgren and Aagaard, 2014; Ritter and Andersen, 2014) and has been the subject of a still growing number of academic and practitioner-oriented studies. The extensive stream of work on business model innovation has generated various important insights. Yet, our understanding of business models remains fragmented as stressed by Zott et al. (2011). One thing all the authors in this field seem to agree on is that a business model is a model of the way in which a business does business (Taran, 2011). However, while there is consensus on the meaning of "doing business," namely creating and delivering value so as to generate value and achieve a sustainable business position, there is less agreement on the "model" part (Taran et al., 2013). Another key challenge of performing studies in BM and BMI relates to the issue addressed by David J. Teece, who states that "*the concept of a business model lacks theoretical grounding in economics or in business studies* (Teece, 2010, p. 174)".

Business models exist in multiple forms. They can be applied as a core unit of analysis extending beyond the business boundaries (e.g., Zott and Amit, 2007). In addition, business models may be viewed as a construct between strategy and implementation (Baden-Fuller and Morgan, 2010). Business models can also be a mean for commercializing new technologies (Chesbrough and Rosenbloom, 2002; Chesbrough, 2007; 2010) and as an intermediary between different innovation actors such as businesses, financiers, and research institutions, that is, actors who shape innovation networks (Doganova and Eyquem-Renault, 2009). Business models can therefore be subject to innovation themselves, or a template for implementing managerial initiatives (Zott and Amit, 2010). Furthermore, they can be used to depict current realities ("as is") or used for simulations to decide on a preferred future ("to be") (Osterwalder, 2004; Chatterjee, 2013; Lindgren, 2013), that is, as role exemplars (Baden-Fuller and Morgan, 2010). Business models (real) can then be seen as a representation of strategic decisions, which have been implemented through tactical choices (Casadesus-Masanell and Ricart, 2010), which may create self-enforcing "virtuous circles" in processes and resources, as stressed by Casadesus-Masanell and Ricart (2011).

Business models can also have a narrative role (Magretta, 2002), serving as boundary objects (Doganova and Eyquem-Renault, 2009) and as conventions (Verstraete and Jouison-Lafitte, 2011) or theories of performative actions (Perkman and Spicer, 2010) in which stakeholders become motivated to participate in the joint realization of a venture. As such, the core idea of the business model concept addresses many classic questions of strategic nature such as market relevance (value proposition), what customers to serve, and how to serve them, how to make a profit, technology (Magretta, 2002; Sandberg, 2002; Morris et al., 2005; Verstraete and Jouison-Lafitte, 2011).

Baden-Fuller and Morgan (2010) underline that from a holistic and systemic concept, a business model perspective may be expected to contribute to a sustainable business model innovation (SBMI) agenda by opening up new approaches to overcoming internal and external barriers. Birkin et al. (2009a, b) identified in their study on North European and Chinese businesses that societal and cultural demands of sustainable development evolve outside the economic sphere as drivers for business model change in businesses. Their findings reveal that as social and natural needs become institutionalized as concrete societal and cultural demands—business model ecosystem requirements (Lindgren et al., 2015), these business models will change radically and businesses are expected to ensure adaptations in order to secure legitimacy, legality as well as business success. The definitions of sustainable business model innovation originate from different scientific areas. Looking into the literature on sustainable entrepreneurship and corporate sustainability management, the concept of sustainable business models is still used in a fuzzy way (Stubbs and Cocklin, 2008; Lüdeke-Freund, 2009; Schaltegger et al., 2012).

However, earlier work reveals the first developments in mapping the concept of and movements toward sustainable business model innovation. Lovins et al. (1999) for one propose a four step agenda to align business practice with environmental needs, which they labeled "Natural Capitalism." The four steps constitute the following: increase in natural resources' productivity; imitation of biological production models; change of business models; and reinvestment in natural capital. What is important for our review and mapping of the concept is the fact that Lovin's and colleagues come across a fundamental change toward sustainable business models as crucial to realizing Natural Capitalism and business potentials in the future.

Another interesting early contribution that emphasizes the same understanding of sustainable business model innovation is Hart and Milstein (1999), who see sustainable development as a force of industrial renewal and progress. They conclude that "simply transplanting business models" from one economy to another will run counter to sustainable development (Hart and Milstein, 1999, p. 29). Common for these two classic articles is how they see changing business models as a way to reduce negative social and ecological impacts as well as a way to achieve sustainable development.

More recent scientific contributions mapping the sustainable business model concept reveal a more elaborate understanding of the components involved. For example, Yunus et al. (2010) reason that for social businesses to evolve, a specific business model framework is needed that integrates a social profit equation. They present a number of key components, which go into explaining and developing a social business model as presented in their model (Figure 2.1).

According to their concept, social businesses apply business models that above all recover their full costs and pass profits onto customers who shall benefit from low prices, adequate services, and better access to maximize the social profit equation. Yunus et al. (2010) refer to this as follows: *"a no-loss, no-dividend, self-sustaining business that offers goods or services and repays investments to its owners, but whose primary purpose is to serve society and improve the lot of the poor"* (p. 311). Another interesting contribution from Boons and Lüdeke-Freund (2013) in mapping sustainable business models suggests different typologies. They define three different types of sustainable business models, which create social value and maximize social profit focusing on three different areas (pp. 14–15):

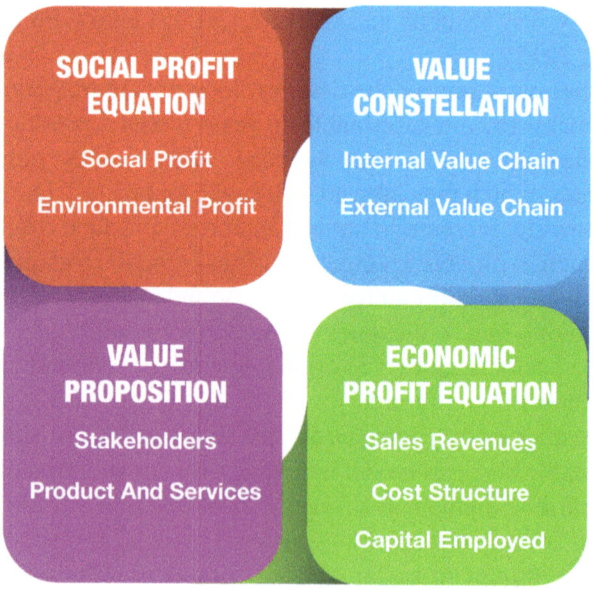

Figure 2.1 Components of a social business model.

Source: Yunus et al. (2010, p. 319).

1. Technological innovation: Creating a fit between technology characteristics and (new) commercialization approaches that both can succeed on given and new markets.
2. Organizational innovation: Implementing alternative paradigms that shape the culture, structure, and routines of organizations and thus change the way of doing business toward sustainable development.
3. Social innovation: That helps in creating and further developing markets for innovations with a social purpose.

Other streams of literature emphasize that the SBMI typology change depending on the kind of partnership (e.g., public–private, business/NGO collaboration), which is required to create social value and maximize social profit (Kanter, 1999; Chesbrough et al., 2006; Dahan et al., 2010; Lodsgård and Aagaard, 2016). Summarizing the different attempts to map SBMI, we draw on the discussion presented by Doganova and Eyquem-Renault (2009) identifying business models as a market device. In understanding sustainable business models as a way to build linkages between actors that are necessary to successfully market a sustainable product or service (Boons and Mendoza, 2010), various elements being open to multiple interpretations may be considered strengths rather than weaknesses. In other words, the so-called fuzziness of the concept of sustainability may actually be a useful quality in developing sustainable innovations (e.g., Tukker and Tischner, 2006; Boons, 2009; Hansen et al., 2009). However, these statements are in sharp contrast with our attempts to map and identify SBMI. We therefore suggest further research on SBMI and of how sustainability is constructed by actors involved in value creation (Boons and Lüdeke-Freund, 2013).

2.4 The Business Benefits of CSR and Sustainable Business

To act sustainably provides key benefits that make CSR interesting from a business perspective. When the company strengthens its relations with the organization's stakeholders, it can prevent and limit potential conflicts in connection with its business activities. The closer dialogue with stakeholders also enables the company to make better decisions based on a deeper understanding of the expectations society has of the company. At the same time, working diligently on communicating sustainability help to improve the company's reputation and stakeholder and public confidence in the organization. Thus, companies can gain competitive advantages and create access to new markets and new innovation opportunities through their approach to CSR and sustainable business.

Risk Management

Especially for large international companies, the improvement of the organization's risk management is a critical factor. This is done through, for example, the incorporation of responsible supply chain management and the integration of "code of conducts" in the organization's various national and global supplier partnerships (Yu, 2008). Through sustainable supply chain management, companies can strengthen their risk management and minimize possible risks in their external collaborations and subcontractors (Walker and McBain, 2008). When the requirements for cooperation are clearly stated and agreed upon, the likelihood of misunderstandings between the buyer company and supplier are reduced. And in case the supplier fails to comply with the requirements, the company can terminate cooperation and demonstrate to society that the requirements were part of an agreement, which the supplier failed to live up to. This does not necessarily take the responsibility and/or blame of the company buying the services, yet it establishes a clearer view of what the partners have agreed to and thereby map the line of responsibility of the partners. Yet, even though the company may have the right on their side, having set up the necessary sustainability requirements, any dispute in which the supplier, for example, has used child labor or underpaid employees will eventually still cause bad PR for the company buying the service. Responsible supply chain management and integrating code of conducts in practice therefore require more than written agreements. Ongoing supervision and dialogue is a necessity to success. Engaging with third world suppliers as a sustainable business therefore implies continuous and targeted development and not just an exchange of goods and services.

Optimizing Resources

Sustainable companies and businesses that have emphasized environmental elements in their CSR strategies and are dedicated to preserving the environment and saving energy may also be able to achieve considerable savings from, for example, increased resource efficiency, lower energy and water consumption and waste, which all constitute economic benefits (Cramer, 2005). Lean is often implemented as part of the company's CSR strategy and lean is proven to be especially effective in reduction of resource consumption and waste across processes and functions (Eriksen et al., 2005). Several companies work with CSR of sheer necessity to find alternative solutions to scarce resources such as oil, water, and/or employees with specific skills (Veleva and Ellenbecker, 2001).

Retention and Healthier Employees

In general, corporate codes of conduct are written statements of principles or policies serving as the expression of a commitment to particular enterprise conduct (Diller, 1999). Labor-related codes of conduct usually specify norms and rules by which to evaluate labor practices at workplace (O'Rourke, 2003). Organizations that prioritize social elements and their employees in their CSR strategy and activities experience improved employee health, better work–life balance, and heightened well-being among their employees. These benefits can again help the company in improved detainment of skilled employees. Furthermore, improved well-being will help reduce illnesses, stress, and depression among the employees, which again reduces absence due to illness, which can cost companies a bundle and represent both monetary and non-monetary business benefits (Mirvis, 2012).

Recruitment and Generation Y

An important benefit of CSR is that the company can strengthen its ability to recruit skilled employees. Many companies strategically apply their CSR and sustainability activities in their corporate branding to portray their values, communicate their raison d'être and to increase interest among new employees, particularly from Generation Y. Generation Y, also known as Generation Why, is the generation born between January 1977 and December 1997 and represents a generation of young, talented, and educated young people entering the labor market in recent years. According to the study by Tapscott (2008), accountability, development opportunities, and "meaning" in the job becomes high when representatives of Generation Y choose between different workplaces. But pressure also comes from within a new type of employees and managers who want to use their professionalism also in a social context. They will not compromise their values or contribute to growth that is being created at the expense of human welfare and environmental safety. They want to make a meaningful difference in their work and not just have a job at some company.

Retaining and Attracting Customers

Furthermore, another business advantage of CSR relates to the ability to retain existing and gain new customers, for example, by opening up and entering into sustainable partnerships with customers, suppliers, NGOs, and other partners that can assist the company in developing their CSR and more, sustainable business. Companies can also benefit from connecting their CSR profile with their products and services as part of their branding and communications strategy, and their R&D strategy where CSR can be geared toward the company's internal and external communications and the development of new innovations (Maon et al., 2006), as discussed later in Chapters 5 and 7. It is argued that superior financial performance is delivered through CSR primarily due to reputation effects (see Orlitzky et al., 2003). Supporting this notion, consumer research argues that a strong CSR image enhances brand differentiation (McWilliams and Siegel, 2011), brand equity (Hoeffler et al., 2010), competitive advantage (Porter and Kramer, 2002), consumer attitudes, purchase intentions, and loyalty (Maignan and Ferrell, 2004).

Attracting Investors

There is an external pressure for change in the form of increased demands from investors who begin to weigh the environmental and social responsibility as key elements in their investments and risk management. A significant amount of risk exists in investing in companies that do not emphasize on sustainability (Diller, 1999). The social media has put immense effort to open up the world to enhance it with transparency. People in a growing number can and will keep track of how companies operate and how they invest in and/or buy products from behave. The business benefits from working with CSR and sustainable business have been considerable. The "business case" argument for CSR, namely the leveraging of recurring CSR undertakings to gain direct financial benefits and to improve long-run firm-level competitiveness in terms of profitability and growth, has often been made to justify such undertakings as a wise investment (Margolis and Walsh, 2003).

However, the motivation for companies to work with CSR comes from several different areas: Some companies are pressed from the outside and have a reactive approach toward CSR and sustainable business and, for example, view CSR more as a risk management. Others are more proactive and consider the needs and wants that sustainability generates as a business opportunity. One example hereof is the financial and industrial group general electric (GE), which has developed a so-called Ecomagination strategy focusing on the development of green technology solutions. Interestingly enough, they do not talk about ethics and morality of their sustainable business—but of the "first-mover" advantages and the opportunities to strengthen their competitive edge.

2.5 Case Example: GE Ecomagination

GE Ecomagination was launched in 2005 as is part of GE's growth strategy of how to enhance resource productivity and reduce environmental impact at a global scale through commercial solutions for customers and through GE's own operations. As a part of the strategy, GE is investing in cleaner technology and business innovation, developing solutions to enable economic growth, while avoiding emissions and reducing water consumption, committing to reduce the environmental impact in their operations, while developing strategic partnerships to solve some of critical environmental challenges at scale. The company has reaped $160 billion from the program since 2005. These revenues grew twice as fast as total company sales, making GE Ecomagination a very lucrative, sustainable business investment. The company committed $10 billion in R&D for "clean tech" technology by 2020 and put at the top of the list of priorities a goal to develop alternative water technologies for fracking[1] natural gas. GE launched the third installment of the Ecomagination Innovation Challenge, an open innovation competition that previously asked the world for ideas to help "power the grid" and "power the home." In those incarnations of the initiative, GE and its venture capital partners invested $140 million into selected clean tech businesses. In an interview made by Andrew Winston for Harvard Business Review (Winston, 2014), GE's Chief Marketing Officer, Beth Comstock, says that the company is spending "record amounts" on renewables-related clean tech, including "energy storage, solar, LED, lighter materials, and advanced manufacturing," as well as energy efficiency, distributed energy, fuel cells, and biofuels. GE's Global Communications Leader, Gary Sheffer, poses the question: "Who's doing more research on renewable energy in the world than GE?" and with the amounts that GE spend on research on renewable energy, they are at the very top of the list. In GE's approach toward sustainable business and sustainable innovation, they are pursuing what Sheffer calls an "ambidextrous" strategy, with the main objective is to make the dirty forms of energy cleaner as well. This dual focus is a central part of Ecomagination, Sheffer says, and key in driving GE's sustainable business initiative.

[1]Fracking is a contentious process, but one that environmentalists have supported as long as there is no significant methane leakage during extraction, which negates the greenhouse gas reduction that comes from substituting gas for coal.

2.6 Critique of CSR and the Bottom Line of Sustainable Business

Both theoreticians and practitioners are speaking for and against CSR and sustainable business. On the theoretical side, it is especially the American economist and Professor Milton Friedman that through a shareholder perspective arguments against CSR as a valid business strategy. He is renowned and often quoted for saying: "The business of business is business" in an article in New York Times Magazine (Friedman, 1970), namely emphasizing that the main task of companies is to maximize profits to owners and shareholders. At the same time, he states that leaders do not have the expertise or social skills to work with CSR and that it dilutes their principal purpose and make companies less competitive globally. In the article, he underlines that companies only need to make sure they create growth and jobs, so they pay the "debt" that they may have to the community. A renowned theoretical advocate of CSR is the American philosopher and Professor Edward Freeman, who through the stakeholder perspective combines the resource-based and the market-based with the political- and social-based approach to running a business (Freeman, 1984). The point of the stakeholder concept is that there are other stakeholders than shareholders, who have an interest in the company. There are many different definitions of who should be incorporated under the term stakeholder. Freeman (1984: 189) has a broad definition of stakeholder as "any group or individuals that can affect or be affected by an organization's goal setting," which actually includes all that affect or are affected by the company and its activities, which in practice means all the stakeholders. The stakeholder concept is a substantial addition to the CSR research, which establishes an indication of who the company in reality is accountable for and thus how and against whom CSR should be measured.

The unsuccessful application of CSR in generating business results is also emphasized in more recent literature. Most Fortune 500 firms have engaged in some form of recurring CSR undertakings though even in these large companies *the majority of corporate contribution programs are diffuse and unfocused* (Porter and Kramer, 2002, p. 58). Porter and Kramer (2006, p. 80) observed that *the prevailing approaches to CSR are so fragmented and so disconnected from business and strategy as to obscure many of the greatest opportunities for companies to benefit society.*

In addition to the theoretical criticism, CSR has also received criticism in practice, in terms of how sustainable business can be competitive and measure the positive results of sustainability on the companies' bottom lines.

Several analyses have been conducted on whether CSR has a measurable and positive effect on companies' bottom line. In a research article by Professor Joshua Margolis from Harvard University and Professor James Walsh of the University of Michigan, 99 international studies from the period 1971 to 2001 were examined in mapping the relationship between earnings and CSR. The analysis showed that 55 of the studies demonstrated a positive correlation, four a negative relation, 22 no association, and 18 an ambiguous relationship (Margolis and Walsh, 2001). The problem with this comparative study is that the studies vary according to the method of operationalization and data collection, which makes it difficult to draw definite conclusions about the link between CSR and the financial bottom line.

Another large international study conducted in 2007 by Goldman Sachs reveals that companies that prioritize CSR has a 25% higher return than companies that ignore it. A study among SMEs in 2005, which was prepared by Harvard University and the Foundation Strategy Group (FSG), indicates a positive correlation between CSR and financial earnings (Harvard University and Foundation Strategy Group 2005). The study showed that there may be positive economic effects of especially four types of CSR activities:

1. Activities aimed at addressing CSR in the company's product.
2. Activities that improve employee conditions.
3. Environmental activities that have a quantifiable positive economic impact on the corporate bottom line, such as reductions in resource consumption.
4. Activities that affect the framework conditions, such as support for the educational institutions from which the company recruits its employees, or participation in the development of a regional business strategy.

There is therefore no unambiguous data on the capacity of sustainable business and CSR to generate positive bottom line results in companies. However, both research and analyses point to a generally positive relationship between CSR and the bottom line, also independent of company size.

An interesting question relates to how does corporate responsibility differ from the sustainability of the actions of customers and consumers? This question was posed by Danish Commerce in the survey "Doing well by doing good." The result shows that 42% of companies surveyed believe that they to a large extent should be accountable to the community (DI, 2008: 11). Accountability is also something you exhibit as a consumer, and consumers indicate in the survey that they are aware of their responsibilities, while 48% believe that the consumers largely should show responsibility.

Yet, while nearly half of consumers report that they should take responsibility, the sales figures for responsible, organic and fair trade products speak another language, as only 7% of the food-retailing trade sales is made up of these products. Consumers apparently do not practice what they preach. However, the report does not discuss the fact that the supply of organic products is significantly lower than of conventionally manufactured products, which may also have resulted in a lower percentage of organic products being purchased.

2.7 Success Factors of Integrating CSR into Business

A large body of CSR literature on CSR integration in business has been devoted to organizational responses to external stakeholder demands. However, there has not been much work on how firms attempt to integrate CSR initiatives in business and as a result achieve internal fit (Wenlong et al., 2011). Researchers usually tend to evaluate CSR initiatives from the perspective of societal stakeholders, with only limited attention devoted to either the difficulties associated with the internal bundling of CSR initiatives and prevailing business practices or to the various paths available to overcome these difficulties. The effectiveness of CSR initiative implementation often depends on linkages with other routines in the organization; an appropriate response to CSR challenges may require close coordination across relevant functions (Westley and Vredenburg, 1996). Inadequate cross-functional coordination and organizational barriers (Cordano and Frieze, 2000) can lead to internal conflicts and ultimately weak performance toward achieving societal and corporate goals. Literature emphasizes a number of factors of importance in successful integration of CSR into business, where the factors of CSR leadership, sense-making, and strategic fit are discussed in the following.

2.7.1 CSR Leadership

From a managerial perspective, adopting new, recurring CSR initiatives can be complex and risky, not only because managers have to decide whether or not to respond to a variety of internal and external stakeholder pressures but also because they have to evaluate whether recurring CSR initiatives will actually fit with currently prevailing practices (Wenlong et al., 2011). Leading academics in the research field of CSR, including Waldman and Siegel (2008), agree that the empirical studies of CSR have ignored the leader's role in the implementation of CSR in business. Instead, we must resort to management literature for answers, which show how different types of management behavior and management communication can affect employee motivation and involvement in the integration of CSR and sustainable business. Walumbwa et al. (2008) found in their study that authentic leadership, where the leader maintains his/her integrity through his personal values, showed a clear link to job performance.

The researchers defined this leadership style as "a pattern of behavior that draws on and promotes both psychological capabilities and a positive ethical climate that supports greater self-knowledge, an internalized moral perspective, a balanced dissemination of information, and relational transparency in relation to managers working with employees while supporting a positive self-development (Walumbwa et al., 2008: 94). Likewise, Uhl-Bien et al. (2007) explain that the leader types who live this leadership style, which he calls "complex leaders," create the future rather than directing it. They use language to create a common understanding of the challenges they face and create space for employees so that they can innovate as individuals and learn as a social group. Both definitions speak of leadership that creates the optimal environment for the employees' positive self-development rather than management and administration of the employees and their behavior.

2.7.2 Sense-Making—When CSR Makes Good Business Sense

Sense-making is stressed in CSR literature as a key factor in successful CSR integration. The concept implies that CSR is applied and explained in relation to a practical context to make sense for the stakeholders involved (Angus-Leppan et al., 2010). In practice, this means that the individual sales person, procurement assistant, marketing coordinator, and/or production worker has to understand what specifically he or she has to do differently in relation to previous (non-sustainable) practices and how they should integrate CSR into their new routines and behaviors. The majority of the extant CSR research has viewed CSR initiatives as a response to the specific demands of external stakeholders (Jenkins, 2005), to enhance corporate reputation (Fombrun, 2005), prevent legal sanctions (Parker, 2002), respond to NGO actions or manage risk (Husted, 2005), and generate customer loyalty (Bhattacharya and Sen, 2004). Such a predominantly external focus of contemporary CSR practices has led to the bulk of these CSR activities being disjointed and "almost never truly strategic" (Porter and Kramer, 2002, p. 57). For CSR to make good business sense, managers have to provide and communicate a clearer link between existing practices and sustainable business and relate sustainable business behavior to the specific functional context of the organization.

2.7.3 Strategic Fit

Senior managers need to assess to what extent integrating CSR into business and adopting new, recurring CSR initiatives as practices might disturb current routines. At minimum, the positive contribution to long-run business performance should outweigh the additional costs resulting from any disturbance to current business practices. Inherent in the discussion of alignment is the notion of "fit" of CSR practices in terms of expected internal coherence, consistency with prevailing business routines and contribution to business performance, thereby focusing less on external consistency with stakeholder demands that is discussed extensively elsewhere (e.g., Atkinson et al., 1997). The concept of "fit" has been a central theme in the strategy literature (Venkatraman and Camillus, 1984; Miller, 1996). It is commonly understood that organizations, as systems of interconnected practices, must achieve a fit both with their external environments (e.g., Lawrence and Lorsch, 1967; Pennings, 1987) and internally across strategy, structure, and processes (e.g., Chandler, 1962). The importance of internal fit has also been highlighted in the CSR literature, such as the suggestion on the impact of "*the consistency between an organization's overall strategy and its CSR activities and [the coherence] within the varieties of CSR activities contemplated during any given period of time* (Basu and Palazzo, 2008, p. 129)".

The above levels of internal fit, when combined with the external fit between CSR initiatives and societal stakeholder demands for particular CSR activities, will ultimately determine the credibility and effectiveness of CSR initiative outcomes.

2.8 Supporting the Quality of Sustainable Business and CSR

CSR, beyond being a strategic management tool, is also a way to support quality, but in a new way. For CSR as a quality tool is not just about ensuring the greatest homogeneity in products and processes or to generate the lowest error rate. The concept of quality takes on a new tone and a different content when integrated with sustainability. Quality in this context is defined more by the perception of the society and by impact that the product quality has on the environment. The discussion concerning quality and quality management's role and usefulness in relation to CSR and sustainable development has lasted for more than a decade (e.g., Garvare and Isaksson, 2001; McAdam and Leonard, 2003; Isaksson, 2006; Klefsjö et al., 2008). Quality management is a management approach that has been suggested to project a number of synergies with sustainable development, in general, and environmental sustainability, in particular. For example, the principles of customer focus and continuous improvement are applicable in practices supporting environmental sustainability, especially in product development.

Such synergies could be used to the advantages of organizations who are faced with the challenges of current era to develop and manufacture products with considerations, not only toward quality, cost, and time but also to the environment. In this concept, environmental sustainability here refers to minimizing or eliminating harmful impacts, such as reduction in wastes in terms of materials, resources, and energy (Daly, 1990; Goodland, 1995) in the development and manufacturing of products and services. Other critical elements to consider in product development are the economic and social sustainability in relation to, for example, the cost of poor quality (Isaksson, 2005) and the social life cycle impact assessment (Dreyer et al., 2006).

Corbett and Klassen (2006) emphasize that "*the quality management revolution has pushed the operations management community to adopt a broader perspective, including processes upstream and downstream, as well as the organizational context surrounding those processes . . . the principles of quality management have become widely accepted in theory and practice*" (p. 9).

Bergman and Klefsjö (2010) define the quality of a product in the following manner:

Its ability to satisfy, or preferably exceed, the needs and expectations of the customers (p. 23).

And with the customer requirements of today, these quality needs chiefly center on sustainability, eco-friendliness, cradle-to-cradle, and proper stakeholder management across the value chain of the product.

2.8.1 ISO Standards as a Tool in Sustainable Business

ISO certifications are an integral part of many companies' quality work today. The International Organization for Standardization (ISO) has developed more than 16,500 international standards targeted several different areas.

A partnership between ISO and the Center for Social Responsibility has developed an ISO guidance standard for sustainability by the name ISO 26000 Social Responsibility. The goal of this guide is to animate all types of organizations to voluntarily work with social responsibility and guide organizations in concepts, definitions, and evaluation methods in the field.

ISO 26000 was originally developed as a general guideline for CSR, but the challenge with the guideline is that it cannot be used as a management system standard in all types of organizations and in accordance with the applicable international agreements and conventions.

Several researchers underline the strength in connecting quality and CSR and point out that quality management can positively contribute to a more effective implementation of CSR (Castka and Balzarova, 2007; Gendron, 2009). At the same time, they warn against the use of quality standards for sustainability as management will risk isolating and standardizing complex CSR issues, which are not suited to be resolved using a standard (Schwartz and Tilling, 2009).

The idea behind ISO 26001 is to provide organizations with the tools to develop a more effective system to fulfill the company's social responsibility. Specifically, the management system standard supports its efforts to do the following:

1. Establish a policy for social responsibility
2. Establish objectives and processes to meet the requirement in the policy
3. Implement the necessary actions to improve performance and
4. Demonstrate and measure that the system meets the requirements of this standard

ISO 26001 is only a tool and a framework for working with CSR, whereas the content to be loaded into this framework and the skills with which the tool is used determine the success in practice. An effective CSR integration like so much else depends on the skills and commitment at all levels of the organization and functions, especially in top management, which has the responsibility to integrate social responsibility into the organization's policies, strategies, and activities, and to act as role models for the behavior change that implementation of sustainable business requires. Expectations, continuous dialogue, and particularly training are the prerequisites for CSR integration success, as neither employees nor managers can be expected to possess the necessary knowledge and skills required to be able to embed CSR in business in a quality-conscious way throughout organization.

2.9 The Ethical and Philanthropic Dimension of Sustainable Business and CSR

Ethics and ethical responsibility are central elements of CSR. In addition, the ethical element of sustainable business is part of what distinguishes CSR in relation to other management and strategic approaches. The previously mentioned CSR pyramid of Archie Carroll includes the general well-known corporate responsibility (the economic and legal responsibility) and the ethical and philanthropic responsibility. The ethical responsibilities do not just incorporate doing what is responsible and right in the economic and legal sense, but also to go beyond the economic and legal requirements and voluntarily strive to be morally responsible in the business choice the company makes.

The ethical responsibility includes both economic and legal responsibilities. It is ethical to create wealth through a business, while it is also perceived as an ethical challenge to follow the laws of society. The ethical responsibility is broader than economic and legal responsibility as it requires more than just profit maximization within the law. Carroll (1979: 500) explains that the ethical responsibility comes from society's expectations of the company, which goes beyond the economic and legal responsibility. Society therefore identifies and determines the content of the company's ethical responsibility. The requirement for the ethical responsibility of society also reflects the ongoing debate and changing values, and this continuous development makes it difficult for company to always meet the expectations and thus the ethical responsibility, as defined by society. It is up to the company to decipher the signals given by the community and thereby shape the company's own values for sustainable business (Carroll, 1999).

The fourth responsibility type in the CSR pyramid, the philanthropic responsibility, relates to how the company carries out philanthropy in a way that is beneficial to society. Corporate philanthropy is often closely associated with the efforts to give the company a positive public image in the community. Corporate philanthropy involves reflection on how the company's earning is best used to create mutual benefits for business and society. Central to the philanthropic responsibility is that it goes beyond what is expected of a company. So there is here focus on a voluntary basis of responsibility, not based on any coercion or expectation from society (Carroll and Shabana, 2010).

Corporate philanthropy has long had its stronghold in the United States and the United Kingdom, where large companies every year donate considerable sums of funds and sponsorships for various purposes. The companies that donate the most money are often characterized by economies that equal the BNPs of small countries. Thus, one can talk about these organizations having a very direct and visible economic impact on society due to their size and resource consumption. One of the largest international comparative analyses of philanthropy is conducted at Johns Hopkins University. It compares 36 countries philanthropy from 1995 to 2002. The findings from the study reveal that the United States is a clear number one in the category of voluntary donations, which represents 1.85% of US GDP. Philanthrophy in Ireland and the UK constitute respectively 0.85 and 0.84% of GDP, whereas Sweden, Finland, and Norway are next to each other on the list with 0.35 to 0.40%, while Italy and Germany is at the bottom with 0.11 and 0.13% (Salamon et al., 2003). The study concludes that private philanthropy is closely linked to the tax burden. The Anglo-Saxon countries have historically had a lower tax burden than Germany, Italy, and the Nordic countries. However, if volunteering is included in the comparison, the analysis from Johns Hopkins University reveals that Netherlands and Sweden go straight to the top and past USA (Boettke and Coyne, 2008).

The international differences in the level of business philanthropy also relate to the diversity of social structures. A comparison between philanthropies in, for example, Danish and American companies emphasizes that the difference can be attributed to Denmark's welfare model, which ensures a social safety net with among other free hospitals and free universities, which USA and UK do not offer. There simply has not been the same needs or focus on private corporate philanthropy in Denmark until now. The PR value of philanthropy is today so considerable that companies see it as a beneficial part of their branding and as a symbol of their sustainable values and business integration of CSR. The company's priorities and choice of philanthropy and sponsorships are visible proofs of the company's values. More and more employees prioritize corporate values, public image, and social responsibility high when choosing jobs. At the same time, customers are increasingly demanding sustainable products and companies and investors prioritize suppliers and partners with whom they can develop in the long run also. Philanthropic behavior portrays a surplus, not just economically, but socially and ethically too, which assists companies in creating a sustainable public profile that helps to retain existing and to attract new employees, customers, suppliers, and new partners and stakeholders.

2.9.1 Indulgences and Harmful Products

It is always positive when companies sponsor charities and the poor and needy, but if it becomes a way to buy indulgences, then it suddenly becomes a dangerous strategy. For example, if an oil company that has just dumped oil into the oceans can buy their freedom and "good conscience" by, for example, sponsoring schools in Africa, then the philanthropic aspect of CSR has taken a wrong turn. CSR is about the company's entire social responsibility and sustainable behavior, not only the areas where it suits the company to be sustainable. This also means that if the company's products potentially have a negative effect on the environment or cause other damages to society, then damage control and minimization can be the place to start for a company's sustainable business and CSR. The following two case examples reveal how companies with unsustainable products can be and act sustainable too.

2.10 Case Example: FMC and Cheminova

FMC acquired Cheminova in April of 2015. Pro forma revenue totaled approximately $4.5 billion in 2014. FMC employs approximately 7,000 people throughout the world and operates its businesses in three segments: FMC Agricultural Solutions, FMC Health and Nutrition, and FMC Lithium[2]. FMC and Cheminova are very much aware of the fact that their portfolio of products, counting pesticides, and other products protecting crops from pests, weeds, and disease has an impact on the environment that they have to account for. They therefore chose to focus their efforts on minimizing potential negative environmental effects, for example, by explaining their customers how they dose correctly, minimizing overuse of pesticides, and how they minimize and recycle waste of packaging. In addition, they advise and assist customers in choosing less toxic alternatives, while phasing out the most toxic products in their product portfolio. One could ask the simple question, why do they not just stop their production, and that way ensure less pollution in the world. However, the world is regrettably not that simple. Crops help feed the world's population every day, and if the output is diminished due to pests, weeds, and disease, hunger will be the consequence. Farmers with large productions of, for example, wheat or sugar are expected to produce at affordable and competitive price, and they can only do this, if they have a high yield and efficient production of crops, which pesticides

[2]http://www.fmc.com/agsolutions/Home.aspx

helps to ensure. Thus, the demand for pesticides will always be there, until other and more effective and sustainable solutions are found and integrated. And again, if FMC or Cheminova did not exist, demands would just be satisfied by another supplier, unless the requirements for productivity and costs are changed, or the substances are prohibited by law. If the demand for pesticides should be reduced, then quite different and more complex and long-term mechanisms need to be initiated. It starts with consumers and what they want and are willing to pay for.

The growing demand for organic products reveals a change in consumers' perceptions of a product's value and quality. However, comparing the organic production with the non-organic reveals that the production of organic produce is negligible. Approximately 37 million hectares is now operated as organic farming, which represents about 0.9% of the world's total farmland (Willer and Kilcher, 2011). So if all consumers suddenly only demanded organically produced agricultural products, a global supply problems and lack of space would quickly emerge, because organic farming requires significantly more space, due to the lower yield per km. The CSR solutions to environmentally harmful products such as pesticides are just not so simple, when you think in global terms. However, it is critical that companies, which manufacture and sell toxic and/or dangerous products, think about viable ways to minimize and reduce the damage and negative effects of their products and production on the environment and society.

2.11 Case Example: British Aerospace

Another interesting example of how to minimize the negative effects of dangerous products is seen in arms production and in the development of land mines. The use of mines was banned in 1997 by the Ottawa Convention, which today counts 155 nations, but land mines are still used, and large countries such as the USA, Russia, China, India, and Pakistan have still not signed the Ottawa agreement, which commits them not to use land mines. Thus, total removal of the product is apparently not possible. Therefore, one of the world's largest weapons manufacturers, British Aerospace, BAE, has developed a biodegradable land mine, while traditional land mines remain in the ground and continue to function and harm, even when the war is over, ecological landmines dissolved after five years. That way, you avoid mines causing damage long after a war has ended, which is not an elimination of the problems of land mines, but a reduction.

This way of thinking CSR and sustainable business in the same context as weapons' production and toxic pesticides may well seem very provocative and cross-border for many CSR specialists. The best alternative would obviously be to ban these products by the law, but as long as there is a demand, the harmful and toxic products will be produced and sold by someone in the world. Although land mines are prohibited by the UN Convention, they are still used in wars in the third world. So if nothing else, it is more socially responsible to minimize accidents and dangers caused by the products than to do nothing about the consequences.

3

Sustainable Business as a Strategy

Both researchers and practitioners agree that CSR policies and CSR strategies are important tools to collect and highlight a company's work with social responsibility, just as they also commit the management to act responsibly (Muller, 2006; Russo and Tencati, 2009). Fundamentally, CSR is about behaving properly and socially responsible as a corporation, and the company's CSR strategy and policies describe and explain how the company specifically expects to carry this out in practice. A CSR policy can be described as a formulation of how the company takes on their responsibility for their actions and their impact on society, the environment, and their stakeholders. However, mere creating a policy is not always enough to convey the company's good intentions. It must co-exist with either a strategy or plan, which explains how the company should carry out its intentions through concrete actions. In integrating sustainable development into the strategic planning of a company, sustainability aspects have to be taken into account in the analysis of external developments and internal strengths and weaknesses. A sustainability strategy integrates the social and environmental dimension into the strategic management process of a company (Baumgartner and Ebner, 2010). A CSR strategy can be applied as the company's guideline of how to work with and integrate CSR across both the internal and external organizations. Different CSR strategies can be distinguished, ranging from reactive strategies to offensive and proactive strategies. The company's strategy for corporate responsibility can also be interpreted as simultaneous progression and as categorical models (Dyllick and Hockerts, 2000; Holmberg and Robèrt, 2000; Baumgartner and Ebner, 2010). Baumgartner (2014) suggests the following categorizations of CSR strategies:

- Introverted—risk mitigation strategy: Focus on legal and other external standards concerning environmental and social aspects in order to avoid risks for the company.

- Extroverted—legitimating strategy: Focus on external relationships, license to operate.
- Conservative—efficiency strategy: Focus on eco-efficiency and cleaner production.
- Visionary—holistic sustainability strategy: Focus on sustainability issues within all business activities; competitive advantages are derived from differentiation and innovation, offering customers and stakeholders' unique advantages.

Strategic planning is often based on the classic method of forecasting (Dortmans, 2005) where opportunities and risks that result from external developments and strengths and weaknesses of the company are assessed and future developments are anticipated. Another approach is *backcasting*, which is based on the idea of first defining a desired future state and afterward planning strategies and actions to achieve this desired state (Holmberg and Robèrt, 2000). Quist and Vergragt (2006, p. 1028) define backcasting as follows:

First creating a desirable (sustainable) future vision or normative scenario, followed by looking back at how this desirable future could be achieved, before defining and planning follow-up activities and developing strategies leading towards that desirable future.

In the following sections, we will look at the various ways of integrating CSR into business, implicit and explicit CSR strategies, a CSR integration model, and how to report on sustainable business and CSR strategies.

3.1 Strategic CSR—Integrating CSR into Business

Relatively little research has been carried out within the area of the actual management, implementation, and the operational side of CSR where operational CSR is mostly discussed in terms of a project management structure (Castka et al., 2004; Sachs et al., 2006). In this book, a coherent design of CSR implementation, integrating CSR both horizontally and vertically, is suggested. In this context, horizontal CSR integration is understood as integration of respectively internal and external CSR. In many organizations, CSR is characterized by practices, which have been aggregated into CSR bundles. These CSR bundles have in some cases been translated by companies into CSR strategies depending on the level of systematic approach and alignment with the existing business strategy. The study by Aagaard & Lemmergaard (2016) argues that in order to facilitate a vertical and a diagonal integration, it is necessary to uncover the level of consciousness and system building in the field of CSR. Researchers have given special attention to the link between CSR and company performance (Waddock and Graces, 1997; Stanwick and Stanwick, 1998; Lopez et al., 2007; Bird et al., 2007; Weber, 2008; McWilliams and Siegel, 2011). However, what is missing is a discussion on how CSR communication and activities are integrated horizontally. Without the horizontal integration at both the internal and external levels, companies are missing important performance opportunities (Aagaard & Lemmergaard, 2016).

Internal CSR integration is understood as the level of employee involvement and the content and the effects of internal oriented CSR communication and activities. In explaining the internal CSR integration, the employee engagement pyramid (Melcrum, 2006) can be adopted, which explains the move from of CSR awareness to understanding and believing in the CSR strategy with the ultimate goal of employees committing to act in a social responsible way. Moreover, internal CSR integration includes the social topics of engagement in fair and efficient HRM, guaranteeing safety, occupational health and security, respect of freedom of association, abandoning of discrimination, and encouragement of diversity (Martinuzzi et al., 2010). External CSR integration constitutes the external CSR communication and involvement of customers, suppliers, and other stakeholders in the externally oriented CSR communication and activities of the companies.

Vertical integration is understood as a holistic approach toward CSR integrating the external and the internal CSR operations. According to the list of generic CSR topics, as developed by Martinuzzi et al. (2010), this includes:

- Environmental topics (i.e., clean air and water, biodiversity, minimizing toxic substances, emissions, sewage and waste, conserving natural resources, applying renewable energy and avoiding the usage of raw materials, engaging in climate protection, facilitating reusability, and recyclability of products)
- Global topics (i.e., raising stakeholders' awareness for social and environmental topics, practicing sound stakeholder management, facilitating sustainable supply chains, respecting human rights, engaging in poverty reduction, and participating in the development of public policies)
- Social topic (i.e., respecting consumer interests)
- Economic topics (i.e., pursuing sound corporate governance practices, ensuring transparency through economic, social, and environmental reporting, engaging in fair competition, combating bribery and corruption, fostering sustainable consumption and production).

A further dimension is added to the vertical integration, namely the diagonal integration, which should be understood as strategic CSR. This dimension leans on Burke and Logsdon's (1996) strategic CSR understanding. From this perspective, the following concepts are relevant:

1. Centrality (i.e., fit between CSR policy or program and mission and objectives of the organization)
2. Specificity (i.e., the organizations ability to capture and internalize the benefits of a CSR program)
3. Proactivity (i.e., in absence of crisis conditions)
4. Voluntarism (i.e., discretionary decision-making and absence of externally imposed compliance requirements)
5. Visibility (i.e., observability of a business activity and the ability to gain recognition form stakeholders)

3.2 Implicit/Explicit—Informal/Formal CSR Strategies

The fact that some companies do not have an explicit CSR policy is of course not synonymous with the companies not working with CSR. However, the more implicit CSR approach and the lack of external communication can backfire. For it is not only politicians and the media that emphasizes CSR. Also, employees, customers, and stakeholders stress the necessity of sustainable business (Dawkins, 2005; Panapanaan, 2006; Sones and Grantham, 2009). The company's commitment to CSR can be a way to signal its social responsibilities clearer to the outside, which can improve the company's brand and the attraction of talented employees and loyal customers and investors. Prioritizing CSR in corporate strategy and business has become a business premise of today and a symbol of good corporate practice and a company's license to operate.

While corporate social responsibility (CSR) is becoming a mainstream strategic concept to consider for many organizations, most of the research to date addresses CSR in large businesses, and limited research has been performed on the relationship between small- and medium-sized enterprises (SMEs) and CSR. Recent research stresses the need for re-orientation away from the large multi-national firm as the primary focus of CSR and business ethics research (Thompson and Smith,1991; Tilley, 2000; Spencer and Rutherfoord, 2003; Grayson, 2004). Even though the social debate and research primarily focuses on large companies' CSR, the SMEs also feel the pressure. The requirements from, for example, consumers, customers, investors, and the companies that the SMEs supply bring about a need for CSR integration into business. Particularly, if the SME wants to qualify as a suppliers, a partner and for investments. Although many SMEs are socially responsible and work in a sustainable fashion, it may be necessary to communicate more explicitly how the SME conveys its social commitment to the outside world.

The interest and motivation for starting to work strategically with CSR must exist inside the organization, and particularly in the management group, as they will have to prioritize and allocate resources to the area. However, a pressure from outside may also motivate the company even more to pursue sustainable business. For example, when employees and management are asked about their (lack of) CSR activity or lack of CSR reporting by an existing or potential customer or when a competitor decides to integrate and communicate a CSR strategy and its actions for sustainability. Most SMEs can report several sustainable initiatives, but often lack a systematic approach, and need to provide an overview and collect initiatives, and link its CSR activities to the business of the company. Working diligently with strategic CSR entails, among others the application of CSR policies and code of conducts, which makes it possible for the company to gather and communicate their CSR focus to the public.

Yet, there are several considerations that companies have to make before they prepare and integrate a CSR strategy. More and more scientists underline that there are differences in how CSR is carried out in large international companies and SMEs (Perrini, 2006; Morsing and Perrini, 2009) and that as an owner or manager of an SME, it is wise to consider whether an informal or formal CSR strategy is the right approach, as discussed earlier in this chapter.

Other considerations in the integration of CSR strategies relate to whether the company should choose to develop an overall/global CSR strategy or local CSR strategies if the company has more than one unit and/or has globally dispersed divisions. Alan Muller (2006) argues that a global strategy can help ensure effective transformation of CSR practice across the organization. However, global strategies may lead to a lack of ownership and legitimacy at the local level. Conversely, a local or regional strategy can be much more responsive to local characteristics, yet in turn become too fragmented compared to the overall strategy and policies of the parent company.

Many SMEs have not identified their CSR practices or only sporadically, which often means that only the manager or business owner is aware of what actually makes for sustainable initiatives in the company. In addition, smaller organizations rarely have the opportunity and/or can afford to hire specialists to carry out CSR. Therefore, it is often a task that the owner or one person in, for example, HR or Communications is made responsible for.

Several researchers indicate the importance of distinguishing between informal and formal CSR strategies among SMEs (Grayson, 2004; Perrini, 2006; Russo and Tencati, 2009). The informal and formal CSR strategies have different purposes, which means that the two strategy types should also be applied differently. Where the informal CSR strategies can help to support the daily work with CSR, it in turn misses the strategic aim and the involvement of external stakeholders. The informal strategies have their eligibility in small organizations where everyone knows everyone and where the practical side of working with CSR and sustainable business takes on a central role. Some of the disadvantages of the informal strategies include that the company rarely reap the rewards of sustainable business in the form of more business, new customers, partners, investors, and qualified employees, because the company though an informal approach ignores to communicate and brand itself on sustainability.

The formal strategies are important tools of communication, both internally and externally, and can assist in ensuring uniform practices, better coupling of business, stakeholders, and CSR and the harvesting of the financial results and opportunities generated through the communication of wellperformed, sustainable business. The fact that CSR strategies are formal and explicit is not synonymous with the integration work being performed faster or easier. Where the informal strategies are already a part of everyday life in the organization, the formal strategies have to be incorporated and made an explicit part of practice and routines of the company. Oftentimes, the strategic development is represented as a continuum—going from informal to formal strategies—as the company over time develops a more professional and structured approach toward sustainable business (Aagaard & Lemmergaard, 2016).

3.3 A Model for Integrating Sustainable Business

In this section, The CSR house is presented as a model tool for developing and integrating sustainable business and CSR strategies in practice. The model is developed based on theoretical and empirical CSR research (including Werre, 2003; Cramer, 2005; Morsing and Vallentin, 2006; Pedersen and Neergaard, 2007; Urip, 2010). The model is generic, which also implies that each step should be tailored to the unique context, characteristics, and situation of the organization. This model is not exhaustive of all the considerations that an organization should make about the integration of sustainable business, but it provides a guideline for the generic steps and activities that management should perform to establish the appropriate knowledge base for more effective development and integration of sustainable business. The CSR house presented in Figure 3.1 is a further development of Cramer's (2005: 586) integration model, which consists of the following phases: (1) List the stakeholders' expectations and requirements; (2) Formulate CSR vision, mission, and code of conduct; (3) Develop CSR strategies and action plans. (4) Set up a monitoring and reporting system; (5) Integrate process quality and management systems; (6) Communicate the results achieved. Cramer's

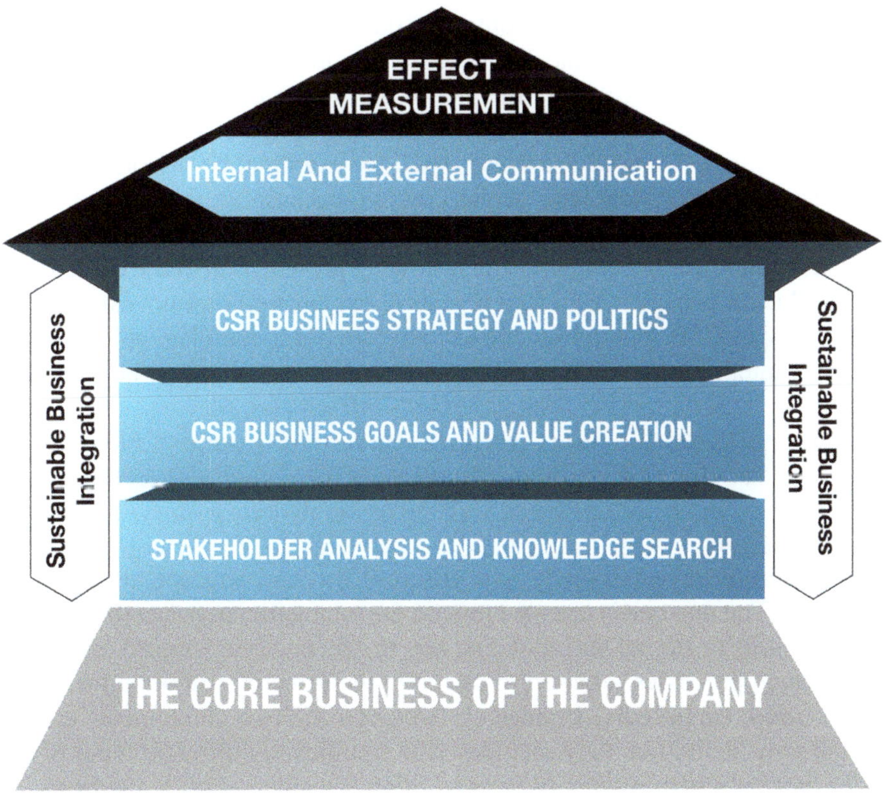

Figure 3.1 The CSR house.

Source: Inspired by Cramer (2005, 586).

model as well as The CSR house emphasize the operational side to CSR implementation and stress the integration of sustainable business across the company while emphasizing the necessity of linking CSR to the company's existing business.

The CSR house constitutes the following six phases in sustainable business integration:

1. The core business of the company
2. Stakeholder analysis and knowledge search
3. CSR business goals and value creation
4. CSR business strategy and politics
5. Internal and external communication
6. Effect measurement and evaluation

The six phases are generic, but the content of the activities carried out during each phase should be targeted and tailored to the company and its unique characteristics and situation. Although the phases are presented chronologically, this does not mean that the practical implementation of the phases follows same chronological order. Some of the phases will, for example, be performed in parallel and overlap or start later in the process, because the company has already implemented earlier stages. Content and duration of the phases is also dependent on the specific company. The following presentation of the content of each of the six phases is therefore a general explanation that provides an overall review of CSR integration and is therefore not a complete guideline for practical integration of CSR in all types of companies and their business strategies.

The green paper by the European Commission in July 2001 defines CSR as: *a concept by which firms integrate the principles of social and environmental responsibility in their operations as well as in the way they interact with their stakeholders.*

This definition reveals clearly two different ways to conceptualize the phenomenon: The first one is oriented toward the interaction between the firm and its (primarily) external stakeholders, and the second one focused on the internal change processes required to integrate the principles into the firm's operations (and, supposedly, strategies). Although the two components are part of the same definition, they each describe significantly different types of activities, not only with respect of their location (outside vs inside the firm) but also, and primarily, with respect to the distance to the core activities of the firm. These perspectives are presented by the lateral phase of "sustainable business integration" and are integrated in the model from phase 1 to 6 as explained in the following.

3.3.1 Phase 1: The Core Business of the Company

Initially, the steering committee and/or the management team has to be selected for the assignment of integrating sustainable business. Management representatives of each company function should be selected for the committee/group, to ensure that value creation through sustainable business is optimized across the company's value chain. The composition of the steering committee should be made with respect to the purpose, objective, and vision for sustainable business and CSR in the company, and should support the width and depth of how the integration process should be carried out.

First, the steering committee/management group should reconcile why the company wants to operationalize sustainability into their business and across their functions, products, and practices. Key questions to ask in this alignment phase of the integration process could be the following:

- What does the company expect to get out of working with sustainable business?
- What is the "true" purpose and objectives of integrating sustainable business across the internal and external organization?
- Which role should sustainable business play in the company's present and future?

Too many companies skip this basic step and mistakenly believe that everyone in the management team agree on the objective, the output, and the integration of sustainable business. If the management group does not agree on the above, then they will automatically end up communicating it differently to their part of the organization, which quickly can undermine the success of the rest of the integration process and confuse employees unnecessarily (Aagaard, 2012).

When the objective, purpose, and expectations are clarified, then the next part has to address and assess how CSR is linked to the company's existing business and practices, and how CSR can create value (both monetary and non-monetary sustainable value). The interconnections among these practices may lead to internal/external fit/misfit, thereby enhancing or reducing business performance (Siggelkow, 2001). It is therefore imperative that business and sustainability strategies, practices, and activities are developed together and aligned and merged in the most optimal fashion. Relevant questions to ask could be the following:

- In what way can our CSR practices actively support our business strategy, vision, and mission statements?
- How does sustainable business fit with the company's existing products, processes, and services?

- In what way does the company want to generate social and sustainable value?

The purpose of this section is to highlight and discuss the relationship between the company, the business, and CSR, to ensure that the CSR activities identified are supporting the business and sustainable growth, and in ways that can be explained to employees, business partners, and other stakeholders. In practice, the committee/group should review the company's business strategy and assess how the implementation of sustainable business across the value chain can assist in achieving the business strategy and goals of the company (Aagaard, 2012).

3.3.2 Phase 2: Stakeholder Analysis and Knowledge Search

In ensuring optimal CSR integration, knowledge has to be sought internally and externally and through the company's key stakeholders in order for the organization to select and prioritize the appropriate sustainability focus in their sustainable business strategy. Internal and external knowledge has to be obtained about current sustainable and non-sustainable practices of the company and among other key players/competitors on the market. Furthermore, the demands and requirements of customers, investors, collaborators, and other stakeholders in relation to the company's sustainable business have to be identified and aligned (Miles et al., 2006). This knowledge search and stakeholder analysis can consist of the following activities:

1) *Internal, sustainable, and non-sustainable business practices*

   The first step is to map the existing CSR activities and sustainable business of the company in relations to what the company is already doing on sustainability. Many companies have already launched various initiatives supporting social and environmental responsibility in different ways, but they may not apply the sustainability terminology in public when communicating these activities. Furthermore, the initiatives may not be aligned with corporate strategy and/or be structured or systematic in ways that ensure sustainable business performance. A small brainstorm about what the company is actually already doing can create an overview of the existing CSR activities. Oftentimes, a CSR strategy or plan can function as a kind of "umbrella" for the existing CSR-related practices that may have had other labels, but contain sustainable elements. Such CSR-related initiatives are, for example, related to the inclusive labor market, environmental requirements, and green procurement policies.

Another part of this mapping process requires identifying non-sustainable and potentially harmful activities that the company performs and that it should attempt to minimize or remove in the pursuit for sustainable business performance.

2) *External, sustainable business practices*

Society identifies a standard for what sustainability is today based on what types of sustainable businesses are already on the market. This also means that the average sustainable business performance of competing companies set the bar for where a organization as a minimum has to go in pursuit of corporate responsibility and sustainable business. This does not imply that companies and organizations should copy each other's CSR strategies and -initiatives. Yet, it does not hurt to know what the key players on the market or in the community do to live up to the applicable and current sustainability standards. A competitive mapping of existing sustainable business among competitors and key partners as an indicator of what the company has to aim for is one way to go about it. Another approach is to go beyond existing sustainable business practices and the expectations of society and that way build a new, sustainable business and gain the publicity and competitive first-mover advantages that follow.

3) *Stakeholder analysis and stakeholder dialogue*

Stakeholders can be defined as:

> "*Those groups and individuals who can affect or be affected by the actions connected to value creation and trade*" (Freeman et al., 2010, p. 9).

Two fundamentally different approaches to stakeholder management are identified in literature (Freeman et al., 2007): a *management of stakeholders* approach and *a management for stakeholders* approach. The management of stakeholders approach is derived from the resource-based view of an organization (Pfeffer and Salancik, 1978). Stakeholders are seen as providers of resources. In the other approach, all stakeholders are perceived to have the right and legitimacy to receive managerial attention (Julian et al., 2008). In other words, the stakeholders are not seen as means to the organization's aims, but valuable in their own rights. In this approach, stakeholders are "persons or groups with legitimate interests in procedural and/or substantive aspects of corporate activity." Stakeholders with high harm potential and/or high help potential should receive more attention than other stakeholders (Aagaard et al., 2016).

The stakeholder analysis can be applied as a tool to ensure optimal linkage between the company's sustainable business and its stakeholders. A company does not live in isolation, but is dependent on cooperation with customers, suppliers, and other business partners and is dependent on that these and other stakeholders have a good impression of the company. At the same time, there is an interdependent relationship between a company and its stakeholders, which means that strategic business initiatives will affect the company's stakeholders and therefore should be aligned with them to ensure optimal integration and performance of the company's CSR across their value chain. Therefore, stakeholders should be involved from the start to ensure that the right initiatives are prioritized and that the company involves them in the business opportunities that CSR initiatives can potentially open up.

Maignan et al. (2005) suggests six different groups of stakeholders:

1. Employees
2. Customers
3. Suppliers
4. Shareholders
5. The environment
6. Local community

To this, one may add authorities and other business partners that are not suppliers. Most stakeholder groups are known by the company, but it can be beneficial to review the company's entire value chain from start to finish to also map the stakeholders that the company might not be in contact with (that often), but who has an influence on or are influenced by the company. In practice, stakeholder analysis often takes the form of structured or semi-structured dialogues or interviews with relevant representatives of stakeholder groups. Interview guidelines are prepared to ensure that the company gets answers to the questions and areas, where they seek knowledge. The interview sessions should be reported and can then be applied in the development of the organization's sustainable business.

3.3.3 Phase 3: Sustainable Business Goals and Value Creation

A large majority of the existing CSR literature emphasizes that in order to improve value creation, there is a need for companies to apply a more holistic approach through integration of various functional areas into the CSR interface of sustainable business (Leigh and Waddock, 2006; Van Tulder and Van der Zwart, 2006; Aagaard, 2012). Sustainable business creates value to existing and new employees, customers, business partners, and other stakeholders, and this value creation should be communicated. It is especially important to convey what benefits and value the company's sustainable business specifically provides the new employee, the customer, different partners, and other stakeholder (Louche, 2010). The business case potential and value creation through CSR has been claimed in more business areas, such as:

1. Increased reputation, legitimacy, and image (Weber, 2008; Carroll and Shabana, 2010; Walters and Anagnostopoulos, 2012)
2. Employee motivation, retention, and recruitment (Austin, 2000; Weber, 2008; McCallum et al., 2013)
3. Cost savings, increased sales, and risk reduction (Weber, 2008)
4. New business opportunities and attracting green investors (Carroll and Shabana, 2010)

The knowledge search and stakeholder analysis performed earlier in phase 2 will provide knowledge of what value creation the stakeholders expect. Hereafter, the business goals can be aligned with these expectations and the company's business strategy to ensure a more optimal strategic fit. These sustainable business goals can target any area of relevance to the corporate strategy, vision, and objectives emphasizing everything from improved reputation, employee recruitment, energy consumption to waste reduction, etc. Some companies choose a schematic approach, where the demands, requirements, and expectations of every stakeholder or stakeholder group are stated, and where the benefits and the value that the company's sustainable business can create for this party and the metrics applied in evaluation of performance are presented (Morimoto et al., 2005). The results from the continuous analysis can be applied to see the evolution of the value that the company generates for its stakeholders. Furthermore, it can be communicated both internally and externally to build loyalty and create common understanding and joint development. Even though it is easier to communicate quantitative results, the qualitative results may have an equally or bigger impact, if communicated using positive story telling. It is therefore important that the evaluation of the value creation is both quantitative and qualitative.

3.3.4 Phase 4: Sustainable Business Strategy and Policies

Strategic management is the process of planning, implementing, and evaluating company-wide decision-making, enabling an organization to achieve its long-term objectives (David, 1989). The organization's objectives are specified, and policies and plans are designed to achieve these objectives. Strategic management provides the overall direction for corporate activities based on the vision and mission defined on the level of normative management. In the integration of sustainable development into the strategic planning of a company, sustainability aspects have to be taken into account in the analysis of external developments and internal strengths and weaknesses (Baumgartner, 2014). Ph.D. scholar Christine Hemingway at Aston Business School (2005) stresses that CSR does not only operate by an external pressure, it can also occur as a result of personal morality inspired by managers and/or the employees' own socially oriented, personal values. If the company has witnessed employee-driven CSR initiates, these should be integrated into the strategy and supported to give the employees the continuous motivation for further sustainable business efforts. Based on the business strategy, the core of the business (phase 1), the stakeholder analysis (phase 2), and the sustainable business goals and value creation (phase 3), an overview of prioritized strategic focus areas of the company's CSR and sustainable business (as requested from the internal and external organizations and stakeholders) should be identified. The strategic focus areas can emphasize different, key areas of attention for the company (e.g., health, security, environment, climate) depending on the industry, characteristics, and situation of the company. From these sustainable business focus areas, a number of selected CSR-activities are identified, which should be carried out across the organization and its functions to ensure the achievement of the business goals and the value creation as requested internally and externally. For a comprehensive corporate sustainability strategy, it is necessary to consider all sustainability dimensions, their impacts, and their interrelations (Baumgartner and Ebner, 2010).

Since the CSR strategy is a strategic tool for management's sustainability work, the CSR strategy should adequately explain what the company wants to achieve with sustainable business and the strategic fit, which exists with the company's vision, goals, strategy, and the stakeholder requirements (Urip, 2010). In practice, this will entail that management describes how the company's CSR and sustainable business is directly linked to the overall business strategy and they requirements by stakeholder and the society. In addition, management must explain the choices and prioritizations made in setting the sustainable business strategy and goals, the short-term and long-term initiatives and effects, and how the company expects to evaluate and measure the sustainable effects and business results of the strategic focus areas and CSR-initiatives. Last but not least, management should present a concrete plan that specifies the cross-functional activities and specific activities for each function in the company. Ven Bert and Jeurissen (2005) underline that the more companies integrate CSR in their business strategy, the better companies will be able to satisfy stakeholders' requirements and needs. However, a CSR study by Deloitte's on CSR integration in Danish firms (Deloitte, 2011) revealed that a large number of Danish companies have not managed to embed CSR in their business and overall business strategy. From the study, it appears that only a quarter of companies surveyed said that their CSR strategy was largely rooted in the overall business strategy. The remaining companies were divided into two groups of equal size. One group that only to a low degree have incorporated CSR into their business strategy, and another group, which to some extent has tried to get CSR integrated into the overall business strategy. The results attest to a willingness to work with CSR, but the integration of CSR into the company's business is far from optimal. The survey shows that 70% only have integrated CSR to some or a low level, indicating that there is a development potential for companies to better anchor CSR in corporate business, while making their social responsibility more business-driven.

Embedding CSR also takes place at different levels, giving rise to particular activities and needs, for sustainable business to function optimally. Management should in their implementation planning to reflect on how CSR is anchored best among others, their Board of Directors, across business and the specific company functions and out among the partners of the company (Werre, 2003). In practice, the strategy development and integration process may consist of different integration processes and sessions with different groups of participants. In some cases, top management and the Communications department can assemble and prepare the CSR strategy, which is then presented to the functional areas and their directors and managers, who then integrate the strategy into their operations. In other cases, top management identifies the strategic focus areas of the CSR strategy and the functional managers map function-specific action plans. The latter approach allows for a wider functional integration of sustainable business and CSR. This is due to the functional managers and their knowledge being actively involved in the strategy development process and not just in the implementation process (Hemingway and Maclagan, 2004).

Sustainable business is often implemented through various policies that specify the desired sustainable behavior—internally and externally. Code of conducts is a common name for these policies, which guides the proper, sustainable behavior among employees, suppliers, and other partners. The main purpose of codes of ethics are to assist managers in avoiding hazards associated with immoral actions (Rosthorn, 2000) while reaping the rewards that emanate from moving toward a moral ideal (Garcia-Marza, 2005). For some companies, it may seem tedious having to write down what management think is good behavior in the company, as that they may think every employee and/or supplier know or should know this. However, the fact is often that employees and partners do not necessarily have the same perception of sustainable behavior as management and that "good behavior" can be defined and understood in several different ways depending on the perspective you choose to take.

The process of specifying, writing, and communicating these policies is therefore necessary to ensure a uniform understanding and practice of what is the desired and socially responsible behavior in the organization across the value chain and among its collaborations. A written and public guideline for good behavior often helps to reinforce the seriousness and the resonance that the policies allow the organization. For if management writes it down, then it must really mean something! However, the fact that policies are written is not a guarantee for good integration. Policies and code of conducts have to be explained and understood by those, who have to apply them, so that they make sense for the individual and can be integrated into the daily routines. The development of a code of ethical conduct should create dialogue that values contributions of all involved parties and recognizes the overlapping and competing interests and actions that lead to inevitable conflicts over resource allocations (Hill et al., 2007).

One thing is to implement a policy in an organization, where you know the culture. It is another thing to introduce a policy to a supplier or partner of a different nationality or culture. In these contexts, policy integration will require a greater emphasis on implementation and explanation. In addition, the company can benefit from setting goals and carry out measurements and possibly impose sanctions and consequences if the specific code of conducts is not followed by the supplier and/or partner. In continuation hereof, Schwartz (2002) also stresses the importance of participation in the development and operationalization of moral standards for healthy ethical climates in businesses.

3.3.5 Phase 5: Internal and External Communication

Companies that have worked with corporate sustainability for years are often adept at communicating their CSR work and using it in their branding, recruitment, and business collaborations. However, sometimes, companies are better at communicating externally about their sustainable business than they are at integrating it within the internal organization (Aagaard and Lemmergaard, 2011). Morsing et al. (2008) highlights the importance of employees as a key stakeholder group. In line with this, for example, Dawkins (2005) and Nielsen and Thomsen (2009), suggest that CSR communication should be developed with an "*inside-out approach*" so that the starting point is ensuring employee commitment. Employees are critical in communicating the company's CSR and sustainable business, because other stakeholders see them as a credible information source, and they can, therefore, be useful for enhancing a company's reputation (Dawkins, 2005). Yet, employees are not always involved in decision-making, and internal CSR communication is often one-way communication about decisions made elsewhere in the organization (Ligeti and Oravecz, 2009). By not involving employees in CSR communication, companies fail to utilize the full potential of employees as active CSR communicators and ambassadors (Welch and Jackson, 2007). Different elements of successful internal communication are suggested in literature. For example, Barrett (2002) emphasizes the importance of face-to-face communication to reach employees instead of relying on indirect channels such as electronic media. Vaaland and Heide (2008) underline the centrality of channels that encourage bottom-up communication. In addition, Welch and Jackson (2007) claim that employees should be differentiated based on, for example, demographics or structural levels rather than be treated as a single public to ensure that the information is targeted and is relevant and meaningful for them (Barrett, 2002).

According to Dawkins (2005), communicating CSR requires careful listening to all stakeholders, and then utilizing the information received to operate in a transparent manner to ensure that the key stakeholders understand how the organization operates. In managing stakeholder communication, the company first has to identify and prioritize the stakeholders to be able to analyze their strategic importance to the firm (Cornelissen, 2004; O'Riordan and Fairbrass, 2008) and the action that should be taken (Mitchell et al., 1997).

An emerging stream of research examines how organizations use communication and projected images to highlight their commitment to CSR (e.g., Brammer and Pavellin, 2004; Dawkins and Ngunjiri, 2008; Tata and Prasad,

2015). Highhouse et al. (2009) reviewed the literature on corporate reputation and developed a model that described how organizational investments and other factors act as cues to influence various images in the minds of their audiences. A KPMG (2013a) survey reveals that reporting about CSR actions is a mainstream practice worldwide, undertaken by more than 70% of the 4,100 large companies KPMG surveyed in 2013. This development is an improvement compared to the year 2011 when about 2/3 of the case companies communicated CSR through reports. This finding is supported by a research report prepared by the MIT Sloan Management Review and Boston Consulting Group showing that sustainability communication has increased in the past 4 years. In addition, the study revealed that corporate sustainability is increasingly measured through clear key performance indicators and firmly put on the top management agenda, as stated by 65% of the case companies in the 2014 survey compared to 46% in 2010 (Kiron et al., 2015).

A key communication vehicle for enhancing corporate image is social reporting (Hess, 1999). The CSR strategy report is an important medium for the development and communication of corporate social responsibility. In the debate on CSR reporting, the Global Reporting Initiative (GRI) is a central player as presented in Chapter 1. GRI have produced and offered different frameworks for reporting, which are widely used and accepted around the world. GRI's reporting framework shows the key principles and performance indicators that organizations can apply to measure and report their economic, environmental, and social performance through the triple bottom line. In addition, the CSR strategy can help to increase the focus on and interest in sustainable business internally and externally, as it can be applied as a communication platform for the company's CSR activities addressing all stakeholders, and thereby helping to strengthen society's confidence in the company, while underlining the company's "license to operate."

3.3.6 Phase 6: Effect Measurement and Evaluation

CSR activities have to be followed up and evaluated continuously to ensure optimal value creation as well as the intended sustainable business effects. This follow-up routine has to be integrated into the existing business evaluation systems to help the management and the organization to stay focused, and ensure the continued commitment of employees and partners. The evaluation of the company's CSR strategy and sustainable business performance can be handled as a formal and planned evaluation with disseminations quarterly, semi-annually, or annually and/or as an ongoing and informal assessment of sustainable business taking place at the board of directors meetings, department meetings, partner meetings, and so on. The informal and the formal evaluation approaches have their strengths and weaknesses. Thus a combination is the ideal evaluation approach.

The communication of the evaluation is just as important as the follow-up activity (Urip, 2010). For if the employees have no knowledge of the progress, the successes and the effects generated by the CSR strategy, then they will eventually lose interest and focus. The dissemination of the results can take various forms. Several companies choose to write about the CSR status in employee newsletters through the company intranet and through storytelling, where positive 'CSR stories' are brought forward to the rest of the employees/colleagues, for example, by management.

Impact measurement is an important element of focused and strategic CSR efforts, and the CSR strategy, policies, and goals as defined in the earlier phases play a central role in this process. For the clearer and more explicit the CSR strategy, policies, and activities are formulated, and the more specific and focused the CSR objectives are, the easier it is to make a meaningful impact evaluation and measurement (Cramer, 2005). The effect measurement ensures obviously some control over development, but it should not be used as a rigid control device. Conversely, the effect measurement helps to provide an indication of how the company is performing across the CSR focus areas and across internal and external organizations (Hopkins, 2005). Through the impact measurement, the company can identify the areas that need more attention and/or resources. Morimoto et al. (2005) presents a matrix to evaluate the company's CSR performance, which maps each stakeholder and assesses

Table 3.1 CSR stakeholder matrix

Actor	Environment	Process	Success Factor	Outcome
Private sector				
NGO				
Governments				
Locale inhabitants				
General public				
Suppliers				
Employees				
Customers				
Stakeholders				

Source: Morimoto et al. (2005, p. 321).

the goal of development, the process of the activities, key success factors for achieving the goal, and the result obtained on the basis of the activities undertaken.

The schematic representation of the effect measurement results has several advantages. It makes it easier to communicate, collect data, and to compare the results over the years. However, a schematic representation can never stand alone as a communication devise of sustainable business, because the qualitative results can be hard to communicate in a schematic form. The effect measurement matrix or other schematic overviews of sustainable results should always be supplemented by a written summary that includes both quantitative and qualitative results.

3.4 Case Example: Novo Nordisk

Novo Nordisk started as "Nordisk Insulin Laboratory" in 1923 and is today a world leader in diabetes care. In 2015 Novo Nordisk had over $16 trillion in turnover and employed over 41,600 (March 2016) employees at their facilities in seven countries and 75 subsidiaries and offices worldwide. Novo Nordisk has worked with CSR strategy integration in many different areas and has obtained ownership of the CSR strategy among its internal staff by continuously communicating sustainability by the management group and across the value chain of the company. In addition, the company has established a governance structure, a management system, and "The Novo Nordisk Way," which all guide how all employees should work and operate in a sustainable fashion across the company. The Novo Nordisk Way is an approach all employees and managers have to live and follow, and that is monitored. The three components together make up the CSR framework in Novo Nordisk. However, beyond this framework, Novo Nordisk has also established formal procedures, policies, and guidelines that help to support CSR in daily work processes across the organization. Likewise, Novo Nordisk has developed sustainability goals, which are part of a larger balanced scorecard system, where the objectives are channeled down from division to unity, and down to the individual employee. CSR is also supported through the company's reward and bonus system.

In the HR area, a "people strategy," has been developed, which is based on The Novo Nordisk Way, and that provides the direction for the company's global HR activities in support of Novo Nordisk's strategic goals. Furthermore, Novo Nordisk promotes diversity across the organization by, for example, organizing a global diversity summit for the management group, offering cultural e-learning tools and English teaching to promote collaboration across the global organization, and so on. In aligning HR with CSR, Novo Nordisk has integrated a talent management program (the greenhouse program and the lighthouse program) and employee and management development programs (New department manager's program, the Spotlight program, and the Pitstop program), all of which identify, nurture, train, and develop talents to emphasize sustainable business. Through NovoHealth, the company supports the health and well-being of their employees through sports, healthy food, an individual health check, as well as activities such as the Novo Nordisk Gutenberg Marathon and "We Bike to Work" campaign. Similarly, absence due to illness, employee satisfaction, and employee motivation other key elements, that Novo Nordisk has activities for and measure. In addition, the company has an employee volunteering program, "Take Action", where employees are encouraged to and given the opportunity to participate in voluntary work in continuation of Novo Nordisk's Triple Bottom Line.

In 2010, they donated $10 million to the World Diabetes Foundation and $2.2 million to Novo Nordisk Haemophilia Foundation. Additionally, they have diagnosed 1,300 children with diabetes through their Changing Diabetes in Children program since 2008. They have trained and developed almost 1.2 million professionals in the health services and have trained 494,000 patients with diabetes since 2002.

In relation to the environment, Novo Nordisk has established an environmental strategy, The House of EHS (environment, health, and safety) strategy, where the company through various activities and objectives attempts to reduce its environmental footprint. The company's current strategy is to address the environmental challenges in production through a lean-based approach, the cLEAN program, which focuses on energy efficiency, minimizing water consumption and waste. Novo Nordisk is also working to minimize CO_2 from transport to and from work, business trips, and the transport of all the sales personnel.

All the company's production plants are certified under the international ISO 14001 standard (environment) and OHSAS 18001 (health and safety). In addition, Novo Nordisk applies product stewardship to minimize the environmental impacts of their products throughout the supply chain, and all co-operations and partnerships are regulated by the code of conducts and various specifications for safety, environment, and human rights.

Through an interview with the CSR Vice president of Novo Nordisk, Susanne Stormer, a question was posed in relations to what is the key to success with CSR integration? The response hereto was as follows: *that a company's CSR has to be aligned with the goals between as a company. There must therefore be a strategic fit between the company's social responsibility and business. . . . It should be an integral part of the way we do business.* There is a difference in how Novo Nordisk conducts CSR in global departments, but there is a global system of policies that are equal for all divisions, and the goals are the same and measured the same way to ensure consistency. On the other hand, the specific CSR activities are not only different as there are cultural differences between countries but also within a nationality and business departments that the activities have to accommodate with.

Susanne Stormer emphasizes that *it is key that management supports the CSR activities—before, during and after they are integrated and remember to 'measure the temperature' continuously and follow up.* Furthermore, it is important that management helps to create an infrastructure that ensures the fulfillment of CSR is carried out in the units and the areas where it makes the most sense. This implies that a HR department should take care of the employees' well-being and health, a production department should take care of environmental issues, and procurement should take care of supplier partnerships.

When asked what is the challenge in CSR integration, Susanne Stormer emphasizes: *What is the challenge for many companies is to legitimize it and mind you not only when you can afford and profits. I have put my professional emphasis on connecting CSR to business, because the more legitimate it becomes, and then the easier it will be to disseminate, integrate and measure. . . . It's not what you do but how you do it, you do. How do you make decisions, and how you think the consequences into your actions, are important elements of CSR*".

3.5 CSR Strategy and Sustainable Business Integration

The complete anchoring of sustainable business and strategic CSR can really only be said to be a reality when we no longer use terminologies such as CSR and sustainability or have separate CSR departments to "fix" sustainability. Senior Allan White of the Tellus Institute says that "*paradoxically companies that pursue CSR experience that it becomes less visible as it is integrated—not just in strategy and workflows, but also in corporate governance* (White, 2005: 6)". He describes CSR integration as consisting of three phases:

1. Alignment with corporate objectives and overall business strategy
2. Integration across business units and functional areas
3. Institutionalization of CSR by embedding strategies, policies, processes, and systems in the substance for which the company is made of

Efforts to organize and assemble the sustainability-promoting activities of the organization provide visibility as to what CSR is in the specific company context and increases focus and prioritization thereof in the daily routines and processes.

Practice shows that companies who have worked long and hard with CSR have anchored it so effectively that some choose to phase out their CSR divisions, models and tools, which were necessary in the beginning of their CSR-journey. An example of this is Novo Nordisk, as presented earlier in the chapter. In their case, it appears that sustainability has become an integral part of acting and working, a business premise and a natural part of the business strategy, objectives, and collaborations across the organization. If true anchoring of sustainable business across the value chain and functions is the objective, then employees have to be involved and empowered to work with sustainability, or else sustainable business becomes more of a "publicity stunt" rather than a genuine part of "the way we work around here." At least that is the way it is perceived by employees, who may not understand the connection between the CSR strategy and their own work in everyday life, unless they are actively involved in the CSR integration process (Angus-Leppan et al., 2010).

However, what are the consequences now and in the long run if a company neglects anchoring sustainable business across the internal organization and its different levels and functions? The issue by prioritizing and focusing exclusively on external communications and external CSR activities is that the inadequate performance of the internal sustainability, including employee relations, competence development, and well-being, will ultimately undermine the company and its reputation (Matten and Moon (2008). The main reason is that employees talk to colleagues in and outside the company and industry, and if the word gets around that the sustainable strategy of a company is just 'hot air', then this will eventually affect recruitment, collaborations, and society's perception of the company.

Research and practice emphasize that for CSR to be integrated successfully into business, companies have to establish a cultural interpretation of the

CSR concepts and develop their own CSR approach, which accommodate the cultural differences and the unique characteristics of the organization (Beckmann and Morsing, 2006; Matten and Moon, 2008). However, the majority of the present CSR literature is written by American authors. Thus, to blindly adopt the CSR approaches presented in the American CSR-literature is not optimal, as these models for sustainable business and, management are inherently American, and targeted at American business culture, and therefore do not take other national contexts into consideration. Integration of sustainable business requires a cultural adaption to be successful (Aagaard, 2012). Matten and Moon (2008) reflect in their article on the differences and similarities in the way in which companies in the USA and Europe are working with and communicating CSR to their stakeholders. This discussion has led to a distinction between explicit CSR, which is dominant in the USA, and implicit CSR, which is prevalent in Europe (Beckmann and Morsing, 2006: 119).

According to Matten and Moon (2008), the two concepts are defined as follows:

- **Explicit CSR**—Corporate policies that assume and articulate responsibility for some societal interests: They normally consist of voluntary programs and strategies by corporations that combine social and business value and address issues perceived as being part of the social responsibility of the company (Matten and Moon, 2008: 9).
- **Implicit CSR**—The corporations' role within the wider formal and informal institutions for society's interests and concerns: Implicit CSR normally consists of values, norms, and rules that result in requirements for corporations to address stakeholder issues and that define proper obligations of corporate actors in collective rather than individual terms (Matten and Moon, 2008: 409).

3.6 Organizations Founded on Sustainable Business

Integrating CSR into an existing business is one thing. However, over the past decade increasing attention has been paid to the topic of social entrepreneurs in academic literature (e.g., Austin et al., 2006; Zahra et al., 2009; Nicholls, 2010; Meyskens et al., 2011). Characteristic of these organizations is that they are all based and founded on sustainability. Some of these companies may on the outside look like NGOs in their sustainable emphasis on doing business, but they still have to generate a profit and undertake a business. Their abilities to create social value (Short et al., 2009; Di Domenico et al., 2010; Ruebottom, 2011), quantitatively measure it (Miller and Wesley, 2010), and report it to all key stakeholders, and particularly financiers, are critical skills of social entrepreneurs. These organizations are typically not large multinational companies, but private initiatives, entrepreneurs, and SMEs, which have had corporate social responsibility at the center of their business from day one.

The driving force and the basic idea behind these sustainable organizations and social entrepreneur have in most cases been a lack of sustainable alternatives, which met a specific need or helped solve an ethical, social, or environmental challenge, and where an enthusiast has taken action. For a social entrepreneur, a recognized social need, demand, or market failure usually guarantees a more than sufficient market size (Austin et al., 2006, pp. 6–7). These business initiatives are also often examples of new business innovation, which are made possible through the development of new business models and business understandings that open the company up to new markets and sustainable business opportunities. Examples of well-known social entrepreneurial companies are among others: Terracycle, Toms shoes, Grameen Bank, and so on.

The fundamental ideas of these company founders have not been centered on profit as the primary catalyst for the company's establishment. Oftentimes these business owners are characterized by being passionate about a specific sustainable topic and in having a "mission." Companies can also be based on unique business ideas that make up the "holes" and niches in the market that sustainable business fills with its targeted benefits. An example hereof is Tesla that manufactures electric cars. The demands for sustainable alternatives are there as more and more consumers use their purchasing power deliberately to show their attitudes and personal values.

Social entrepreneurs pursue different entrepreneurial opportunities, which are often related to the social, and so-called third sector markets that tend to be informal, unregulated, and unpredictable, and characterized by the idiosyncrasies of personal relationships (Robinson, 2006). Social entrepreneurs also take different approaches to enact opportunities. For example, social entrepreneurs tend to follow open source approaches or establish loose forms of social franchising (Tracey and Jarvis, 2007; Volery and Hackl, 2010) instead of focusing on privatizing innovation profits by employing patenting. The social entrepreneurship literature particularly highlights the importance of complex relationships and partnerships for social entrepreneurs to succeed

(Austin et al., 2006; Tracey and Phillips, 2007; Bloom and Chatterji, 2009; Zahra et al., 2009). It appears from literature that it is not just only the ability to create social value that is important for social entrepreneurs but also their ability to create a business model that is financially stable and self-sustaining. As such, an ability to incorporate social and financial objectives in one organization is a key element for successful social entrepreneurial (Kistruck and Beamish, 2010).

4

Integrating CSR into Production and Procurement

The concept of sustainable production was first mentioned and discussed at the United Nations Conference on Environment and Development (agenda 21: Programme of Action for Sustainable Development) in Rio de Janeiro, Brazil, in 1992. The conference concluded that the major cause of the continued deterioration of the global environment is due to the unsustainable patterns of consumption and production, particularly in industrialized countries. EU strategies for sustainable consumption and production indicate the direction in which CSR works within the EU. Proposals for new or revised directives on eco-design, integrated pollution prevention and control (IPPC), environmental certification (EMAS) and the EU eco-label, the flower, as well as global initiatives such as Global Compact, are all initiatives to draw companies' CSR activities in the same direction. Sustainable production and procurement is therefore gradually becoming a basic premise for implementing business in the Western world.

The UN commission on sustainable development has determined that the sustainable production and consumption patterns are important cross-cutting issues to be raised at the commission's annual meetings. At the same time, the UN has launched a regional process in which each region has a responsibility to follow up on the Johannesburg Plan's goal of sustainable production and consumption patterns, adopted at the World Summit in Johannesburg in 2002. The Johannesburg Plan invites companies to, among other, ratify and implement international conventions on chemicals and hazardous waste and call for a global chemicals' strategy. During the 2002 Johannesburg meeting, goals were set by 2020 stating that chemicals must be produced and used in ways that do not have significant negative effects on health and environment.

4.1 Sustainable Production

Sustainable production includes the concept of "clean technologies" or referred to as CleanTech where companies are encouraged to conserve resources. The idea of cleaner technology is that, instead of cleaning up environmental problems, companies should prevent pollution by eliminating waste at the source (Danish Technological Institute). Most production techniques can be developed in ways that ensure a more optimal use of natural resources with lower pollution and waste as a result. An active, environmentally oriented products and production policy is important in planning effective consumption of resources in relation to waste, chemical load, the environmental impacts of waste management, and the reduction in non-renewable natural resources.

Sustainable production is explained using the definition of the Lowell Center for Sustainable Production (LCSP), University of Massachusetts, where sustainable production is defined as:

The creation of goods and services using processes and systems that are non-polluting, conserve energy and natural resources, economically viable, safe, and healthy for all employees, communities, and consumers; and socially and creatively rewarding for all workers.

This definition is consistent with the current understanding of sustainable development since it emphasizes environmental, social, and economic aspects of the company's activities. At the same time, the definition is operational in its approach as it highlights the six key components of sustainable production, which companies should optimize as part of their sustainable approach toward production:

- Energy and material consumption (resources)
- Natural environment
- Social justice and community development
- Financial results
- Workers
- Products

To promote a better understanding of sustainable production among businesses, Lowell Center of Sustainable Production (LCSP) defined ten guiding principles (Veleva and Ellen Becker, 2001: 521), which can also be applied by companies as inspiration in their work and the evaluation of sustainable production:

1. The products and packaging are designed to be safe and environmentally friendly throughout their life cycle, and services are designed to be safe and ecologically sound.

2. Services are designed to satisfy human needs and promote equality and justice.
3. The waste and environmentally harmful by-products are reduced, excluded, or recycled continuously.
4. Chemical substances or physical features and circumstances involving danger to human health or the environment are eliminated.
5. Energy and materials are preserved, and the forms of energy and materials are used, which are best suited to the desired goal.
6. The workplace is designed to minimize or eliminate physical, chemical, biological, and ergonomic hazards.
7. Management is committed to an open, democratic process of continuous evaluation and improvement, focused on the long-term financial performance of the company.
8. The work is organized so that it preserves and strengthens the efficiency and creativity among employees.
9. The safety and well-being of all employees is a priority as well as the continuous development of people's talents and abilities.
10. Communities around the workplace are respected and strengthened economically, socially, culturally, and physically, and equality and justice is promoted.

The indicators used by enterprises for sustainable production vary widely. Gallopin (1997) presents a comprehensive analysis of different definitions and indicators for sustainable production and concludes that there is no uniformity in the way the indicators of sustainable production are defined and applied. The failure to identify widely accepted measurement methods and indicators of corporate social responsibility within the production area constitutes a barrier to future efforts of business implementation of sustainable production and business strategies (Ranganathan, 1998: p. 7). The goal should not be to "reinvent the wheel," but rather to build on the work of other organizations such as GRI, Lowell Centre of Sustainable Production, and ISO 14031, which provides various guidelines for the calculation and use of indicators in sustainable production.

There are several important advantages of the company's use of indicators for sustainable production. For example, the evaluation helps to promote organizational learning about sustainable production and can be part of a feedback system that helps managers in daily routines to assess and decide whether they are on track or whether there is need for adjustments. Thus, measurement of the company's sustainable production through a set of indicators becomes part of the company's learning process in social responsibility and in generating sustainable business (DiBella and Nevis, 1998). The specific design of the company's action plans, policies, and indicators for sustainable production must of course be adapted to the unique characteristics that distinguish the company, its objectives, business, and production. In the development of CSR action plans and CSR policies, the specific environmental and resource issues of the company's products, services, and production should be addressed. Selected CSR initiatives should prioritize the areas where company's production has the most comprehensive environmental impact and negative consequences. In practice, this means that each company initially will analyze and assess their current energy, water, and other resource consumption as well as waste and pollution, possibly through benchmarking against similar companies, and then determine which areas are most critical and the targeted interventions that can best optimize the resource consumption and minimize the environmental impact of the company's practice. Lean is one example of a set of management tools and practical processes that many companies apply to improve their company's sustainable production and development.

4.1.1 Cradle-to-Cradle

It is difficult not to mention "cradle-to-cradle", when addressing sustainable production and procurement, as this "mind-set" in many ways determines how the company looks and acts upon the environmental impact of its products and production. When looking at the theoretical development of the area, it appears that the early sustainability initiatives primarily focused on product end-of-life (EoL) strategies such as creating proper product disposal methods, increasing recycling options, and controlling the hazardous waste of after production and emissions from the disposal of products (Sarkis, 1995, 2001; Boks and Tempelman, 1998, Linton, 1999, Chiodo and Boks, 2002). Sun et al. (2003) describe EoL strategies as important decision-making enablers and consisting of planning and integrating product take-back and disposal activities early in the design stage. From these early efforts, environmental requirements have moved upstream to the design stages of products (Johansson, 2002, Kleindorfer et al., 2005, Azapagic et al., 2006).

In elaboration of this, Sarkis (1998) argued that the design for environment (DfE) concept supports integrating environmental requirements into the early design of products.

The cradle-to-grave perspective changed the focus from product EoL environmental impacts to reduced impacts across the life cycle of a product (Rydberg, 1995, Hanssen, 1999, Ljungberg, 2007, Gehin et al., 2008). A product life cycle is defined by Baumann and Tillman (2004, p. 19) as:

the product stages followed from its "cradle", where raw materials are extracted from natural resources, through production and use to its "grave".

Klöpffer (2003) stresses that the systems' approach is necessary since reducing environmental impacts in only one life cycle stage, while ignoring the remaining stages may result in negative contributions to environmental sustainability. Only in this way, trade-offs can be recognized and avoided. Life cycle thinking is the prerequisite of any sound sustainability assessment (p. 134). A practical example of reducing the environmental impact in the "raw material" stage, and ignoring the "end-of-life" stage could be explained using the following example: An engine component of a car may be selected due to the weight and durability of its raw material, contributing to sustainability in terms of low fuel consumption. However, the disposal of the particular material may be hazardous to the environment at the end-of-life stage, which indicates that considerations made at one stage of a product life cycle may result in negative contributions at another stage.

4.1.2 Lean as Part of Sustainable Production

For a manufacturing company, cleaner technology in the form of, for example, recycling and prevention of waste and pollution represents important efforts to achieve sustainable production. Prevention and elimination of waste is a key focus area for lean, and in this context, the waste concept is somewhat broader than just energy and material waste as it also includes, for example, time reductions. In practice, the tools and the focus of management on lean have proven to be effective in reducing waste across processes and functions (DI, 2007a, b, Eriksen et al., 2005). A number of different tools are elementary in successful applications of lean: 5S, value stream mapping (VSM), and single-minute exchange of dies (SMED).

The Kaizen Institute defines the 5S in the following manner:

1. **Sort:** Sort out and separate, which is needed and not needed in the area.
2. **Straighten:** Arrange items that are needed so that they are ready and easy to use. Clearly identify locations for all items so that anyone can find them and return them once the task is completed.
3. **Shine:** Clean the workplace and equipment on a regular basis in order to maintain standards and identify defects.
4. **Standardize:** Revisit the first three of the 5S on a frequent basis and confirm the condition of the workplace using standard procedures.
5. **Sustain:** Keep to the rules to maintain the standard and continue to improve every day.

VSM is a lean management method for analyzing the current state and designing a future state for the series of events that take a product or service from its beginning through to the customer. At Toyota, it is known as "material and information flow mapping" (Rother and Shook, 1999).

SMED is a system for dramatically reducing the time it takes to complete equipment changeovers. The essence of the SMED system is to convert as many changeover steps as possible to "external" (performed while the equipment is running), and to simplify and streamline the remaining steps. The following benefit can be generated through a successful SMED program: lower manufacturing cost, smaller lot sizes, improved responsiveness to customer demand, lower inventory levels, and smoother start-ups (Ferradás and Salonitis, 2013).

Environmental management, characterized by ISO 14001, has been one of the management tools that companies have applied. The practical experiences gained from using environmental management reveal that it is possible to achieve good results, but it requires a serious effort to get there (DI, 2007a). The company's motivation to implement ISO systems often stems from a customer requirement for ISO certification, while the motivation for the introduction of lean often comes from within the company and is deeply rooted among company owners and top management, which can lead to a greater commitment.

Instead of drafting instructions and management manuals, which characterize the way the company work with environmental management through ISO 14001, lean prioritizes employee involvement, for example, by letting self-managed kaizen groups implement the lean tools: 5S, VSM, and SMED and conducting board meetings in the organization. The motivation for sustainable production with reduction in waste, resource overuse, and pollution through lean is ensured through the different tools for employee involvement, which is also why lean in collaboration with sustainable production can be a successful cocktail (Eriksen et al., 2005).

4.2 Responsible and Sustainable Supply Chain Management

A company's supply chain can generally be defined as the number of companies, including suppliers, customers, and logistics providers, who work together to provide a package of goods and services to the end customer. A company's supply chain has economic, environmental, and social consequences for several stakeholders and society, which is also why sustainable supply chain management represents a central pillar of sustainable business. And as stressed by Carbone et al. (2012, p. 477), "*the sustainable supply chain represents a further step towards the institutionalization of a CR strategy*". Sustainable supply chain management also deals with the dialogue that companies engage with their suppliers in order to prevent violations of fundamental human rights and international environmental standards. All in all, the objective of sustainable supply chain management is to live up to the expectations of social and environmental responsibility in accordance with internationally recognized principles and rights, which is defined by the UN. In pursuit of responsible supply chain management, the company has to exceed expectations, working to improve social and environmental conditions in the supply chain. Efforts will typically be aimed at suppliers in developing and emerging countries where regulation and enforcement of laws and conventions is less effective than in most developed countries.

The CSR movement has helped put a strong focus on responsible supply chain management, and many companies find that their customers demand that the company can vouch for the suppliers it uses. This also means that more and more companies are applying their sustainable cooperation as part of their branding as consumers are increasingly aware of and interested in whether the products they buy are produced under ethical and responsible working conditions. The arguments for responsible supplier management in businesses are among others that it can help to improve quality, productivity, and employee conditions at suppliers and helps reduce "scandals" and potential criticism from stakeholders ensuring the company's reputation with customers and partners (Becker-Olsen et al., 2006). The philanthropic rationale for prioritization of sustainable supply chain management relate to the fact that companies through their cooperations can help to improve working and environmental conditions in the supply chain, increasing production and thus earnings and growth in developing countries.

Despite the long history of sustainability, the use of CSR in relation to supply chain management first emerged during the past 15–20 years (Maloni and Brown, 2006). Researchers have tried to generalize the elements that characterize sustainable supply chain management in all industries, and have created concepts such as logistic social responsibility and social responsibility in purchasing (Carter and Jennings, 2002a, b). At the same time, some supply chain researchers emphasize the various themes within a sustainable value chain such as the environment, labor, government procurement, and preferential treatment in procurement (Maloni and Brown, 2006).

The growing interest in sustainable and responsible supply chain management has emerged in the wake of various public counter-attacks against the use of subcontractors, for example, in relation to "sweatshop." Here, consumers have taken an active position and boycotted companies with unsustainable practice exploiting and underpaying subcontractors, particularly in third world countries. However, the real "value" of consumer enthusiasm and commitment to sustainably produced goods can be difficult to assess. Several studies have revealed that consumers are worried about the 'sweatshop' issue and claim that they will pay higher prices to support improvements (Elliott and Freeman, 2000). However, other studies conclude that these studies overestimate consumer willingness to pay extra for the goods in practice (Prasad et al., 2004).

The establishment of the fair trade concept and the fair trade label are results of the growing consumer interest in ethical supplier relations (Krier, 2007). Similar initiatives are constantly arising as consumers increasingly want to have the certainty that the production of the goods they buy have not caused harm to the environment or people. A company has the opportunity to provide consumers and customers with some of the certainty through more targeted and public communication in which they explain and portray through storytelling how they work with sustainable supply chain management, responsible sourcing, and code of conducts. However, the effect of this communication ultimately depends on the relationship that the company has with its customers and consumers, which is built up over time through a sustainable practice and consistent and positive public perceptions in the media.

The company's work with responsible sourcing can receive support from various consulting firms, organizations and government agencies that have developed tools and websites that can assist companies in pursuing responsible supply chain management. A national example hereof is The Danish Ministery of Business and Growth that has developed the knowledge platform: www.csrcompass.com. The CSR Compass is a free online tool that is aimed at small- and medium-sized enterprises in the production, trade, and service industries in assisting them in optimizing their work with responsible sourcing. The tool was originally developed in 2005 and has undergone continuous development. The platform may be aimed at Danish companies, but consist many of the basic considerations that all companies need to make, no matter what country they originate from.

The CSR Compass assigns the following six steps in the start-up and sustainable supply chain management:

1. Reconcile internal ambitions and goals for sustainable supply chain management, and secure internal organization and management support.
2. Conduct a risk assessment internally and in the first, second, and so on, part of the supply chain.
3. Develop requirements for suppliers and a code of conduct with selected issues that are relevant to the company.
4. Evaluate suppliers in relation to your code of conduct and arrange visits.
5. Ensure continuous improvement, explain partnership, and offer knowledge.
6. Communicate efforts and results targeted to relevant stakeholders.

Yet, responsible supply chain management does not only receive praise in business (Nygaard and Aabling, 2011), as the work to ensure accountability and sustainability into the very last part of the value chain can be both impossible in practice and very costly for many (especially smaller) companies. Supply chain management requires knowledge dissemination and training while building the knowledge capacity of the supplier, which can be time-consuming and cumbersome for companies. Particularly for small- and medium-sized enterprises the development and training of the subcontractors' competencies can be a big mouthful. Alternative solutions could be to develop the capabilities of the authorities in the countries at risk so that they are better equipped to enforce local laws in relations to the environment, work environment, and anti-corruption. However, the prospects for such a political process is confusing for even the most determined CSR-oriented companies.

It is therefore suggested that companies start where they can make a difference and focus on their own suppliers and their work with CSR, and through ongoing dialogue, code of conducts, and training, attempt to change the mind-set and the non-sustainable behavior among suppliers that runs against the company's CSR strategy and corporate values.

4.2.1 Code of Conduct—a Tool, Not a Guarantee

Companies that work with responsible sourcing often make use of code of conducts, which in practice really means "guidelines for how we do things around here". Code of conducts consists of documents, guidelines, and toolboxes describing the conditions that the enterprises expect that their suppliers live up to. Code of conducts should be dynamic, and the tools and documents must continuously be developed and adjusted in relation to the changes that occur in the context, conditions and the society that affect the parties. Code of conducts are key for sustainable supply chain management, since they constitute the foundation for the company's creation of a system and procedures, which enables them to continuously assess, whether their suppliers actually live up to the requirements (O'Rourke, 2003).

In the practical preparation of the company's code of conducts, the company should operate from 11 general and company-specific issues, based on a number of international conventions. The items, which companies should highlight in their code of conducts, would need to be aligned to the specific company and its suppliers, cooperation/task, and the conditions prevailing in the countries of both parties. Typical items that the company should highlight in their code of conducts are among others (Diller, 1999; www.csrcompass.com):

1. Forced labor
2. Child labor
3. Discrimination
4. Freedom of association
5. Work
6. Recruitment
7. Use of security forces
8. Management of land
9. Environmental protection and industrial accidents
10. The company's products
11. Corruption and bribery

Points 1–6 constitute general themes that the company's suppliers have direct influence on and which can be directly related to the production of the goods or services that the company buys. By contrast, points 7–11 are the company-specific issues dealing with suppliers' impact on the surrounding community. Thus, the mapping of code of conducts includes an assessment of the elements, which are most important and relevant to the company, its strategy and CSR approach, as well as the situation and the environment in which the suppliers operate. It is a balance between needs and requirements, as the "world" in which the company lives can be very far from the reality that a supplier company operates in.

Existing studies on the social impact of work-related CSR policies and code of conducts performed in various industries (e.g., clothing, toys, sportswear) conclude that code of conducts have potential in the fight against the most immoral and inhumane violations of workers' rights such as child labor, sexual harassment, and corporal punishment. In contrast, these surveys highlight that code of conducts do not provide solutions to problems with low wages, long working hours, and workers' rights to freedom of association and collective bargaining (Yu, 2008). A company should therefore choose its "battles" carefully and prioritize areas for development among its suppliers and subcontractors where the company can contribute most effectively through the introduction of code of conducts.

The dissemination and understanding of the code of conduct among suppliers is crucial for success, which requires more than attaching a code of conduct to the supplier contract. Ongoing dialogues with the supplier about the content and importance of the code of conduct and matching expectations and following practice continuously are imperative. In ensuring that the company requirements are maintained throughout the entire supply chain, the company should ensure and at least ask that their suppliers communicate the company requirements to their subcontractors. However, this is rarely enough and often the company has to conducts some of the dissemination directly to the subcontractors (together with the supplier). Many of the 'media-scandals,' including poor supply chain management, have often been caused by subcontractors' lenient handling of code of conducts. Responsible supplier management should therefore be understood and practiced throughout the entire chain of suppliers and subcontractors.

For even with the most perfect and carefully constructed vendor cooperation agreement and code of conduct agreements, scandals may still occur and will roll out in the press. Thus, a code of conduct is not a guarantee. It only guarantees that both parties (company and supplier) know and have signed what conditions they have agreed that the goods shall be produced under. However, it is not a 100 percent guarantee that these agreements are respected and maintained. Training and communication are therefore key elements to successful and responsible supply chain management, and part of the continuous assessment and dialogue with suppliers. Responsible supplier management is therefore an essential, but often resource-intense task on the downstream companies (O'Rourke, 2003).

In addition to code of conducts, there are also other tools to support the company's socially responsible and environmentally conscious cooperation. For example, the Global Social Compliance Initiative (GSCP) that was launched in 2010 and constitute comprehensive material that companies and their suppliers can apply to define environmental requirements in the supply chain and to ensure implementation of these requirements (www.gscpnet.com). Furthermore, various environmental labels exist, such as the Nordic Ecolabel, the Swan, and the EU Flower, which are useful tools to prioritize environmental efforts in the company's procurement and production.

4.3 Responsible Procurement Management

Responsible procurement management (RPM) focuses on the integration of social and environmental considerations along the entire organizational process aimed at buying inputs useful to the production processes (Carter and Jennings, 2002a, b; 2004; Ciliberti et al., 2008). Procurement plays a key role in CSR, since the choices and the demands that the private and public companies generate affect which products and productions—sustainable and unsustainable—are carried out. The procurement function also has the attention of sustainable business research and practice promoting environmentally conscious and responsible procurement of goods and services among professional buyers—both in public and private companies.

Previous studies suggest that the major barriers to the implementation of RPM are:

1. Costs (Hervani et al., 2005)
2. Resource constraints, for example, limited knowledge, manpower, and skills (Bowen et al., 2001)
3. Power dependencies (Frenkel, 2001)
4. Limited ability to measure the impacts on the correlation to corporate performance (Siltaoja, 2013)
5. The focus on price as the main driver in buying decisions (Krier, 2007)

These findings are supported by empirical data. Analyses show that it is lack of knowledge and not will that is the main reason why sustainable procurement is not prioritized optimally. For example, the market research firm YouGov conducted in April 2011 a representative survey of purchasing managers in Denmark for The Forum of Sustainable Procurement, and in collaboration with Deloitte. The study clearly showed that lack of knowledge and high prices are the main reasons that public and private companies do not procure sustainably. Only half of the surveyed procurement assistants said that they are working with environmental considerations in their procurement. According to the study, the explanation for this was that 30 percent of those interviewed stated that they lacked knowledge. At the same time, 14 percent said that they did not have the support of their leaders to consider sustainability in their procurement. Although the study results were based on a relatively limited basis, consisting of 310 responses, they still give an indication of what is preventing companies from integrating social responsibility and sustainable business to a greater extent in their procurement (Dreyer, 2011).

The challenge of integrating sustainable procurement is perhaps also related to a misconception in the market that sustainable procurement costs more than non-sustainable procurement, which need not be the case. Pursuing sustainability in the procurement function is not about buying the most expensive item of the highest quality, but to assess how the procurement choices that the company makes can affect the environment and the surroundings in the short run and the long run. Examples of daily procurement decisions could constitute how to determine the right, sustainable choice when choosing between different types of light bulbs, printer paper, raw materials, sales rep cars, and so on. The evaluation may be carried out in relation to which alternative is the most sustainable on a short and long run, and what amount and type of waste the company's procurement choice generates Basically, sustainable procurement requires an assessment of the various economic, social, and environmental consequences of buying one product or service over another.

In practice, the integration of CSR in the purchasing function requires a change of the mindset that most procurement departments have typically been rewarded for previously, namely "buy cheap, and get the most (quantity) for the money." This has to be altered for a sustainable mindset emphasizing: "Buy the product that is the most sustainable (economically, socially, and environmentally) in the short and long run." This shift in focus means that procurement choices go from a short-term and cost-focused perspective to address the long-term consequences of the choices that the company makes. However, "best" should not be understood as the highest quality, but as what is best compared to the short- and long-term requirements and objectives that the company has set for its sustainable business and its impact on the environment and society.

However, to avoid procurement from becoming an almost "philosophical" and non-transparent activity, management has to set up clear guidelines as to, what to prioritize the most in making the procurement choices. Furthermore, the procurement guidelines have to be communicated, trained, evaluated, and rewarded in the procurement department to ensure successful integration If, for example, minimization of CO_2, energy consumption, waste, and packaging are central to the company's CSR strategy and sustainable business, then these elements have to be taken into consideration and evaluated before the proper procurement can be made. However, if maximization of ethical relationships with suppliers and/or no child labor is critical to the focus areas of the company's social responsibility, then other choices have to be made. The bottom line is that it quickly becomes complex, as what is good for energy use and waste reduction, may not be good for the quality of the product or ethical working conditions or other scenarios. These choices should be made easier by management through clear guidelines so that the procurement department should not weigh 117 different factors against each other each time they have to purchase a resource, good, and/or service.

The research on RPM reveals that the main enablers for RPM implementation identified are as follows:

1. Corporate values (Giberson et al., 2009)
2. Training (Seuring and Müller, 2008)
3. Collaborations (Vachon and Klassen, 2008) with both NGOs and industry partners

Many companies already have procurement guidelines and procurement policies, but it is not enough to change the procurement requirements on paper. Employees must also understand and be able to see the point of why they need to change entrenched procurement habits. Herein lays an important communication task if procurement is to become responsible in supporting the sustainable business of the company.

4.3.1 Public, Green Procurement

Public companies and organizations have to emphasize sustainable procurement and responsible/sustainable supply chain management at least as much as the private companies. In particular, when you consider the level of procurement that is carried out by each municipality. However, practice at times reveals a different scenario. In the public sector, the focus has mainly been on the so-called green procurement, which involves a deliberate inclusion of environmental considerations in purchasing and tendering. Yet, the majority of public companies and authorities have not come as far in their efforts to integrate ethical and social issues in their supply chain management and in the formulation of their responsible supply chain management (Walker et al., 2008).

Researchers who seek to build a theoretical foundation for public procurement research tend to emphasize such concepts as competitive bidding, principals and agents, transaction costs, and contracting (Flynn and Davis, 2014). However, knowledge of how to implement public, green procurement in practice is lacking (Klay, 2015). Looking at practice, it appears that the private sector in the Western economy for several years has worked actively with implementation of social responsibility across their supply chains (Carter and Jennings, 2002a, 2004; Krier, 2007; Ciliberti et al., 2008). Most production companies are taking a stand today in relation to whether the goods they import are manufactured under ethical conditions where basic human rights, labor rights, and environmental concerns are met. The same attitude and approach should be more widely anchored and structured in public organizations' procurement and at the extent that it could be referred to as sustainable procurement. The difference between public and private companies work with responsible procurement may be due to increased media attention around especially large companies "exploitation" of suppliers in poor countries and the fact that the responsibility is easier to place in a privately owned company than in a municipality, where numerous stakeholders are involved and where the state and government also constitute a central player.

Responsible supply chain management in public organizations is a critical development area for CSR, since the public procurement often represents a large percentage of a country's collective procurement of goods and services. In the future, public procurement must involve considerations of social responsibility in accordance with the conventions underlying the UN Global Compact. The practical challenge of implementing the Global Compact in procurement and supply is, however, that they only consist of general intent and no specific rules, which therefore first have to be operationalized and identified in procurement contracts before they can be integrated and applied (Gürtler, 2009).

5

Integrating CSR into Communications and Sales

Strategic CSR communication is one of the key elements of an effective CSR strategy and in working with sustainable business. If the company fails to communicate its positive CSR initiatives to its stakeholders and the outside world, then reaping the business rewards of the company's sustainability efforts will be impeded. In practice, CSR communication is used to create and develop an organizations' reputation and to give them a "face" that consumers or citizens, partners, and surroundings can relate to (Beckmann and Morsing, 2006).

Furthermore, CSR can help raise an organization's legitimacy and reputation in the eyes of external stakeholders, which can lead to higher sales, lower costs, as well as long-term sustainability (Lam and Khare, 2010, p. 3). CSR is consequently described as a key antecedent of corporate reputation due to its ability to develop core competencies and competitive advantage that are hard to imitate (Melo and Garrido-Morgado, 2012). However, this is only possible when CSR is fully integrated within the corporate reputation (Melo and Garrido-Morgado, 2012). It appears that CSR communication based on the concept of multiple stakeholders and a more holistic and integrated approach to branding is taking hold (Hildebrand et al., 2011). Among others, Simmons (2009) emphasizes that the application of CSR in the reputational area can be leveraged by marketing as a way to attract and keep customers. Consequently, there is an increasing alignment between internal (employer) and external (consumer) branding (Simmons, 2009).

Some might consider CSR as a communication strategy. However, CSR is so much more than just communication although communication is an important focal point in sustainable business integration. The research highlights the many potential business benefits of the internal and external communication of sustainable business and CSR efforts (Maignan et al., 1999). Yet, at the same time, several researchers state that the more the companies mention their ethical and social aspirations, the more likely it is that the initiatives attract critical attention from stakeholders (Ashforth and Gibbs, 1990, Vallentin, 2001).

Thus, CSR communication is a balancing act. Because if the CSR messages are considered to be too "positive" and promise "too much," then the company runs the risk of being accused of 'greenwashing'. Greenwashing is a terminology applied when "green" PR and marketing are applied to deceive consumers by using untrue statements about the company's sustainability in promotion of the perception that a company's policies or products are environmentally friendly even though they are not. On the other hand, an enterprise should not be so careful to tell about its CSR activity that the competitive advantages that they could have gained from a strong CSR performance are lost (Maignan et al., 2005). Even though the Communications department may be very conservative about positive CSR storytelling, they can never eliminate the risk of the company potentially being confronted by the press and stakeholders when it communicates about its accountability. Instead, Communications should develop guidelines and code of conducts, which ensures sustainable Communications and avoids lack of evidence in their positive PR stories. Furthermore, they should establish a "communication emergency plan" in case a PR activity takes a wrong turn in the media.

5.1 Strategic CSR Communication

Several researchers emphasize that the huge communication challenge of companies today relate to how companies strategically communicate about their CSR engagement (Maignan et al., 1999, 2005; Morsing and Schultz, 2006). Morsing and Thyssen (2003: 150) emphasize that companies are in a communication schism since they on the one hand are being urged to engage in CSR initiatives to strengthen the company's reputation, and at the same time, have to be very careful about the way the message is communicated, as stakeholders are critical of too positive stories about companies' corporate social responsibility.

In general, strategic CSR communication can be divided into direct and indirect communications, depending on the stakeholders' knowledge of and interest in the company. To employees, NGOs, journalists, and politicians, the company should apply a direct communication strategy. However, to the general population and to customers, who do not have the same interest in the company as elite stakeholder, they have to use an indirect communication strategy (Morsing et al., 2004: 7). Direct communication is characterized by a high professional level and is carried out through various communication channels such as the company website, annual reports, employee magazines, and other forms of communication generated internally.

According to Morsing et al. (2004), it is risky to communicate directly to consumers as the company may appear as "smug" and self-congratulatory. The company may instead choose to communicate selected messages to its employees so that they can act as multipliers of the CSR information to the public and to the customers. This way, the company communicates indirectly to consumers through a third party (the employee). In practice, this could be done using employees, who appear in advertisements and brochures to tell why a particular company is a great place to work (e.g., sustainable values, health systems and opportunities for competence development). Some companies choose to communicate to consumers both directly and indirectly through famous role models and/or storytelling about successful CSR activities with NGOs or collaborations with other popular partners, which illustrate the company's CSR priorities.

CSR is an interactive strategy as the company does not live in isolation. Thus, CSR should be incorporated in relation to the interactions and relationships that the company has with its customers, partners, and other stakeholders. The company's CSR approach and CSR communication must of course be based on the company and its values and activities. However, it must also be targeted to the value chain and the context of the stakeholders that the organization wants to communicate to and that they are interacting with. Hence, the company can target its communication about its CSR initiatives in ways that can help to improve, enhance, and support the company's internal and external relations (Maon et al., 2006). In ensuring effective communication on sustainable business and social responsibility, the content, channel, and form of communication must be tailored to the specific audience and stakeholder group, affecting and being affected by the company and its sustainable business. It is therefore advisable to use a stakeholder analysis and a stakeholder map that provides an overview of the various internal and external recipients of communication as well as the type of communication that each stakeholder group should receive and in what specific form and through which communication channel. Research emphasizes a number of key elements in successful CSR communication among others:

1) Fit between company and role in society

Brousseau et al. (2013) underline that firms should define who they really are and then explore ways through which "*the ethics of their role in society can be presented to strengthen brand identity*" (p. 62).

2) Ensure consistency and transparency

Beckman et al. (2009) state that stakeholder perceptions of authenticity are important to the success and acceptance of CSR communication and note that stakeholders use available cues, such as transparency and consistency, to ascertain the extent to which the organization is true to itself.

3) Increase credibility

Erdem and Swait (2004) stress that as the credibility of a brand increases, there is a greater chance of a specific brand being included in the consumer's purchasing decision. Similarly, as the perceived credibility of a company's CSR campaign increases, the more likely the consumers are to express positive purchasing intentions.

4) Easy-to-understand and targeted communication

Hämäläinen and Maula (2006) emphasize the need to express a company's CSR strategy in an easy-to-understand format. Furthermore, Barrett (2002) stresses the need to tailor the information to the audience.

5) Prepare for CSR communication crises

As underlined by Tixier (2003, p. 71), *Corporate social responsibility . . . is today, more than ever, an important stake for communication. Companies must know how to communicate if a social or environmental crisis occurs.*

6) Risky businesses require more CSR communication

Organizations from industry sectors with high environmental impact have to respond more to external pressures and communicate social responsibility even more effectively. This is due to the fact that these organizations have a higher level of visibility and their actions are scrutinized by the media, advocacy groups and the public (O' Dwyer, 2003).

Generally speaking, the discipline of CSR communication is probably the part of CSR that is most well described in literature, and it has also received extensive attention in the large enterprises as the CSR department is often placed as part of or in very close cooperation with the Communications department. It is recommended that companies work diligently to integrate their CSR communication with its internal and external communications and in a way that supports the company's core business and value chain. That way the company's overall strategic communications create a more homogeneous and reliable picture of the company, its vision, values, and its business in a natural context of sustainability.

5.2 Internal CSR Communications

The focus of CSR communication, both theoretically and empirically, has centered on the external communication to/with customers and other external stakeholders. However, too much emphasis on external communication can contribute to the perception of CSR being a Communications strategy rather than an integral part of the business strategy and organizational practices of the company. According to Uusi-Rauva and Nurkka (2010) a forgotten core link in CSR relates to one of the company's key stakeholder groups: Employees. A company's employees are of critical importance in CSR communication due to the fact that other stakeholders see them as credible information sources, and they are therefore also useful in enhancing a company's reputation (Dawkins, 2005). Yet, in practice, employees are often not involved in strategic decision-making on corporate sustainability and instead receive one-way messages about decisions made elsewhere in the organization (Ligeti and Oravecz, 2009). In so doing, companies fail to utilize the full potential of employees as active CSR communicators. CSR can only achieve its full potential as a strategy if the strategy is integrated, embedded, and "lived" throughout the organization (Dawkins, 2005). Employees therefore need to be involved and have ownership of the CSR initiatives that the company wants to implement.

The ability to communicate why employees need to spend time on sustainability, what the company get out of CSR, and what the individual's role is in working with CSR is central to anchoring sustainable business optimally across the organization. Juholin (2004) suggests that organizing unofficial meetings such as "Friday coffees" can be very effective venues for communicating strategy. By engaging the employees in CSR business integration, the CSR strategy and practices are continually reviewed, assessed, and adjusted to fit into the internal and external organization. In effective CSR communication, Uusi-Rauva and Nurkka (2010) stress on:

1. The importance of tailoring environmental messages to different employee groups based on what is relevant to them in their jobs
2. CSR messages that might best encourage employees to take environmental action if the messages are clear, practical, and easy to implement, and suggest
3. Utilizing environmentally active employees as internal communicators to promote environmental activity throughout the organization, potentially assigning (environmental) contact persons to each department.

Alignment between the internal and external communications is therefore critical to ensure that what is communicated externally and is also linked to the internal organization. In continuation of this, it is important to remember that the external CSR communication is also read and received internally by the employees. Thus, it is important that employees agree with the external signals sent and do not get "surprising news" from customers or other partners about their own organizations' actions of sustainability. This point is especially true for the interconnectivity between the Communications department and the sales staff. Particularly as this is where potential communication problem can occur. Sales and communication are interdependent, so what is promised in the external communication generated by Communications has to be delivered by the sales department. The continuous dialogue and cooperation across the two departments is therefore important, especially to achieve the full potential and economic benefits of sustainable business.

5.3 Communicating CSR Through Social Media

The Internet has been recognized as a relevant channel for supporting and fostering relations between companies and the public (Waters et al., 2009). Consequently, online channels are becoming an important tool used to manage corporations' reputations (Jones, 2009; Nair, 2011). Social media is a central part of a company's CSR communication, and today, an increasing part of corporate reputation is generated and affected by virtual networks and communities. Overall, social media can be defined as:

The production, consumption and exchange of information across platforms for social interaction (Dutot, 2013, p. 55).

The special characteristics of these media are that they allow a high degree of complexity of viral dynamics, interactivity, and co-production of messages and opinions. Social media are inherently different from traditional media, because they embody two-way communication with asymmetric response (Lyon and Montgomery, 2013). Any generation can be linked with these tools, but it is obvious that companies are attempting to encourage Generation Y customers (between 15 and 25 years old) to interact with them on social media as they would with their friends and families (Lichy and Kachour, 2014).

A report from the research company Forrester Research (2008) documented how the application of Web 2.0 tools such as widgets, wikis, and blogs are growing rapidly at the expense of the traditional offline communications (Jones, 2009). Communication via the company's website and annual CSR reports is often perceived as subjective and a kind of advertisement for the company as it comes directly from the source. On the other hand, the open media allows that other stakeholders and outsiders can give their unfiltered opinion about the company, which seems more objective and credible. This viewpoint is supported by Lyon and Montgomery (2013), who stress that social media reduce the incidence of corporate "greenwash."

According to Dutot et al. (2016) the most important advantages of online social networks constitute the following:

1. An affinity with the target market
2. An ability to increase the brand's reputation with reduced costs
3. The capacity to segment the public and offer real-time metrics

Unfortunately, only few companies understand how to use social media to communicate social responsibility. Rather than seizing the opportunity for dialogues with users, companies end up using the old-fashioned one-way communication from the company to its stakeholders. This is the conclusion of a 3-year research project "*Responsible Business in the Blogosphere*" performed by Copenhagen Business School (CBS). The report stresses that neither the Danish nor the American companies understand how to exploit the two-way communication that social media suggests. Businesses are accustomed to communicating about CSR through CSR reports and advertisements where they will not be contradicted. Yet, if you are present in social media, the rules of communication change. Here, companies can very quickly be challenged and get overwhelming by criticism from angry consumers. Very few companies today are able to handle this, as pointed out by the Professor Mette Morsing from Copenhagen Business School, who is also the leader of the project.

However, social media has a life of its own. The CBS research project uses the example of Apple, which did not predict that consumers would mobilize a "Green my Macintosh" online supported campaign demanding that Apple lived up to the values of environmental sustainability, which the company had promised. The research project also provides another example of Arla Foods, which did not predict a melamine scandal in its Chinese joint venture, although bloggers online had discussed and warned about melamine in Chinese milk products in the weeks leading up to the publication. The CBS research project concludes that online communication has become not only an influential and independent source of reputational development and value creation for companies but also an unpredictable source that is difficult to control. According to Stenger (2014, p. 52), *online presence is a fantastic tool for peer-to-peer recommendation and viral marketing, but it can also damage a brand image or a company's reputation.*

5.4 Supporting Sustainable Business Across Marketing and Sales

The literature on CSR in sales is very limited, which may seem a bit surprising, considering the fact that sales representatives are the ones who must convey the company's CSR and sustainable business approach in their dialogue with the customer. Accountability and corporate social responsibility are elements that are increasingly demanded when entering into partnerships and customer and supplier collaborations. Thus, CSR can be an important selling point to attract, retain, and develop customers and partners in the short and long term. Therefore, CSR should be included explicitly in the company's sales and marketing strategy and in the sales processes of the company's sales representatives. Customers will typically emphasize cost in sales negotiations, it is therefore important to translate the sustainable elements of the company's products into tangible benefits and savings for the customer. A practical example hereof is the international company Grundfos that is a world leader in water pumps. The company has established a sales campaign, "*Meet the energy challenge*," which explains customers how Grundfos' pumps can assist the customers' business in becoming more sustainable and providing savings to the customers by reducing water and energy consumption.

When Sales and Marketing build bridges between the departments, creating joint communication and sales activities are based on the company's CSR strategy, then sustainable business can generate tangible, financial results. In practice, Sales and Marketing representative can jointly prepare sales kits with relevant sales and communication materials that can assist the sales reps in communicating/selling how the company's CSR initiatives contribute explicitly in creating results for the customers while strengthening the customer's business. Furthermore, the Marketing department can identify and develop various CSR sales activities, which can help to support the sales reps in their sales process. An example hereof could be product samples of more sustainable alternatives, phasing out environmentally harmful product alternatives in the customer's product portfolio, collection and recycling of reusable packaging, as well as customer seminars emphasizing joint sustainable behavior and responsible supply chain management.

Basically, the CSR-related marketing of the company and its products and services should be linked to what the sales representatives of the company communicate to the existing and potential customers and users. Worst-case scenario would be for the company to implement a CSR campaign that the sales reps have not been informed of and barely understand themselves in relations to how it can be applied in their own sales processes and CRM. It is therefore imperative that the Sales and Marketing department have a close dialogue about the CSR-based, joint activities to ensure the optimal business effect. Another important benefit is that sales representative can collect customer responses and report back to the Marketing department about how customers, users and partners perceive the external CSR communication and the CSR initiatives launched outside the company. In most private organizations, the sales reps represent the external part of the organization, and they can therefore play a more central role in communicating and anchoring the company's sustainable business externally.

In selling sustainability, the nonprofit organization, Businesses for Social Responsibility (BSR, 2015), suggests that companies should consider the following key areas:

1. **Offer consumers more value from sustainability**: Too often, campaigns focus on what consumers can do for sustainability and not the other way around. Consumers need a tangible value proposition that motivates purchases or actions.
2. **Build functional, emotional, and social benefits**: It is recommended to perform a careful analysis of the barriers associated with the behavior the company wants to promote among its customers, and then support this development and mindset of the customers with a strong value proposition based on functional, emotional, and social benefits.
3. **Pay attention to timing**: Sending consumers the right message, at the right place and at the right time, is an important aspect of sustainable marketing, and perhaps the one area that we know the least about. As mobile and Web technologies become increasingly sophisticated, marketers will be able to fine-tune messaging according to the time of day and receptivity.

5.5 Supporting CSR in Sales and the Sales Department

A CSR-oriented sales approach can help create sustainable development for and with the customers and users. In addition, carrying out the sales process while incorporating sustainable business can provide long-term competitive advantages for both the company and the customers. Thus, integrating CSR into sales should be an obvious part of key account management (KAM). The finest task of the key account managers is therefore to support customer development through the sales process, the knowledge and the professional advices provided. The central idea behind KAM is that the sales representative also acts as a supervisor and advisor for the customer. The supervisor role in

KAM is important for the successful integration of sustainable business in sales and the sales process. Sales reps can actively apply CSR in the sales process, for example, by modifying the existing product portfolio of the customer in ways that reduce CO_2 emissions, energy and resource consumption as well as environmental and ethical issues in the supply chain of the customer, while ensuring the customer specific economic and environmental benefits.

In the previous case example of FSA and Cheminova, the concept of "*product stewardship*" is underlined as a central part of the company's sustainable sales strategy. In the case of FSA and Cheminova, the focus of sustainable sales is on identifying the products that deliver lower toxic emissions while phasing out the most toxic product alternatives to non- or less toxic products, and by providing professional counseling on the optimal handling of the products and optimized recycling of packaging. Thus, important tasks in relation to CSR-oriented sales are among others to continuously overlook the customer's product portfolio, the customer's CSR strategy and its requirements of sustainable supply chain management, while ensuring an optimal fit between the two companies' strategies.

Effective integration of CSR in the company's Sales department and sales processes requires that the sales management is actively involved in mapping the CSR-related sales politics and sustainable sales activities. Some of the key questions the sales department should ask in planning sustainable sales could include the following (Aagaard, 2012):

1. How can a sustainable sale be identified and what does it constitute?
2. How can the company's CSR strategy and activities be applied actively to enhance sales?
3. Which specific benefits can the company's CSR approach provide to the customers and users (e.g., in relations to reductions in CO_2 emission, water and energy consumption, waste and packaging)?
4. How does the company's sustainable products and sustainable sales approach assist our customers in attaining their CSR strategy?
5. In what way can the company ensure continuous development and growth among its customers using sustainable business?
6. How should the company evaluate and communicate the sustainable effects and savings that is generated for the customers?
7. Which types of partnerships and collaborations, for example, NGOs, research organization, universities, and other stakeholders, should the company engage in to support sustainability in sales and in the customers' procurement?

Basically, the sales reps should disseminate how the CSR activities of the company can generate specific benefits and savings for the customer. This implies that the sales personnel need access and knowledge of their customers' CSR strategy to identify the highest level of value creation for the customers, the company's business, and for society. Based on this knowledge, sales activities can be planned to best support new sales, customer retainment, and in building customer loyalty, and co-creation and development with customers.

It is one thing to apply CSR as a unique selling point, but another to practice sustainable management in the sales department itself. The sales department should also facilitate sustainability and accountability in their every day practices and routines. Examples of CSR-related activities in the sales department could, for example, constitute the reduction of the CO_2 emission generated by the vehicle fleet of the Sales division by changing to more fuel-efficient cars or electric cars. Other examples include printing sales materials on recycled paper, donating unsold products or auctioned them way to benefit a specific cause, NGO or local activity. Also, the sales department can provide specific feedback to R&D and production on which sustainable elements in the product or service that the customer value the most and that way identify unmet needs for sustainability on the market.

5.6 Case Example: Hummel

Hummel was established in 1923 and has a long history of creating sportswear with different hummel collections across sport, lifestyle, fashion, and footwear. Throughout the years, hummel has initiated and supported several projects focusing on changing the world through sports by providing sponsorships to poor and war-torn countries. For hummel, sport is more than a physical activity. It is a universal language with the strength to eliminate differences that are facilitated by politics, culture, religion, and belief systems. Hummel works with a quadro-win, which incorporates that the company sets goals for value creation among its four key stakeholder groups:

1. *The Surroundings*: We must think in terms of the effect we have on the outside world, and select cases that we believe in so that we can build our values and brand up on sustainable action.
2. *Employees*: Employees must have a feeling of ownership and suggest how we can do better as the employees to choose a value identity community with the company.

3. *Customers*: We must ensure the value for our customers, suppliers, and other partners.
4. *The company's bottom line*: We must develop and sell innovative and sustainable products to generate positive results.

Hummel is dedicated toward their value of *Company Karma* as the guiding philosophy of how they do business, and it has become the company's version of a more holistic approach toward sustainable business. Company Karma is equal to the company's global commitment to responsible business practices and helps to define the way the company should work with people and the environment. Consequently, hummel has developed a number of guidelines and policies that they follow in their efforts to be a modern responsible company. As a core element of hummel's value system, the company considers their engagement in human well-being to be a key responsibility. This engagement is exemplified through their "Change the World through Sport" projects carried out externally and internally with their employees and supply chain partners. In hummel's pursuit to ensure that their supply chain partners meet the internationally recognized environment, quality, and labor standards, the company endeavors to work with factories that have certifications e.g., SA8000, BSCI, ISO 14001, ISO 9001. Hummel has established a three-part approach toward supplier management, which upholds suppliers to the company's minimum requirements, consisting of

- Code of Conduct
- Internal audits
- Third party audits.

Furthermore, to ensure that the products are free from hazardous chemicals, the suppliers of hummel commit to comply with the "hummel chemical restrictions", which are based on the OEKO-TEX 100; Product class I & II, and REACH standards. REACH is the European legislation regarding chemicals in products and production processes. To ensure compliance, hummel conduct random chemical tests among its suppliers.

Hummel has emphasized CSR as a basic value that should permeate the entire company and be initiated by the employees. The CEO explains, "*one day my sales manager came up with this idea of linking his old boys soccer team with the Red Cross, and thereby we set an event that generated a lot of positive PR and a lot of happy people—including a happy sales manager.*" *CSR must make sense and be a value that permeates the organization.*

And you have to communicate why you're doing this—both internally and externally. It is important to anchor the CSR initiatives in the company's core competencies and create internal and external relevance. "We sponsor football—because we are a sports company!," states the CEO.

Hummel has engaged in several partnerships and CSR-related activities—also some of more innovative character. For example, they collaborated with the American pop band, Black Eyed Peas, on their The E.N.D World Tour 2010 with a Recycling Program. They established a collaboration with the clothing manufacturing company, RETHINK, and the Black Eyed Peas in establishing a unique recycling-initiative focused on producing a sustainable fabric from recycled plastic bottles. Based on this collaboration, recycled T-shirts were manufactured and sold to fans at the merchandise stands at the concert. At the same time volunteers at the Black Eyed Peas tour's wore the T-shirt at all 19 concerts, encouraging fans to recycle their plastic bottles. The same fabric has also been applied in the "hummel recycled line" of sports products while emphasizing hummel's core values: Sport, design, and sustainability. Furthermore, the sporting goods were also launched in Sweden for the men's World Cup in handball.

Hummel is part of the cooperation, Thornico, that constitute a number of different companies, among others: A real estate company, a shipping company firma, and chicken farms. Through the real estate company, a so-called green multi-storey car park was built in Rotterdam. The car park is covered in plants that absorb the CO_2 of the cars in the car park. Apart from these CSR-activities, hummel has also sponsored the national soccer team of Afghanistan and Sierra Leone. Furthermore, the company owns a charity shipping vessel in collaboration with Red Cross; and they own chicken farms in Malawi that are established to support the development of the local community.

The sponsorship of the soccer team in Afghanistan was initially established because Hummel believes in sports as a way to build bridges between cultures. In addition, they also want the people of Afghanistan to have the opportunity to think of something more positive than Taliban, death, and destruction. The specific sponsorship includes sports clothing and equipment, and hummel also raises money for the team to get training facilities in Kabul. In addition, hummel have also put their situation on the public agenda through different types of PR. As an example hereof, BBC showed the soccer match between the American and the Afghanistan women's national soccer teams.

Another core focus of hummel's sustainability is to create a better work place with less stress, higher employee satisfaction, and loyalty, and to attract and retain skilled employees. This is among others created through a strong company culture and a strong brand. The high level of employee satisfaction has a direct impact on productivity, efficiency, and creativity as well as reductions in sick leave. As stressed by the CEO, *CSR creates a higher level of security, and the more peace of mind, and the less anxiety, the more creativity.*

When asked, what is a sustainable leader? The CEO of hummel, Christian Stadil replies:

This can be done in many ways, but the basics must be in order. That is to say that we should treat our employees and our partners, the environment and our stakeholders properly, and in the process, we have to focus and prioritize! How do we do it best? How are we sustainable in the best possible way? Do we need to run 'stop-smoking' courses or provide our employees with more holiday? And/or should we support homeless people or Africa?

Employee involvement and ownership is the alpha and omega—without it you will not get people involved, and it will just be a hollow project. Ask them: What do you think we should do? But it is management that sets the emotional tone, and the leader must lead the way, as stressed by the CEO, Christian Stadil.

6

Integrating CSR in HRM and Administration

The overlap between CSR and HRM is growing, as more and more interdependences are being established through the CSR focus on employees. On the one hand, HRM is promoted to increasingly leverage CSR, while supporting social responsibility across core HRM practices. On the other hand, CSR practices are considered increasingly reliant of HRM for increased effectiveness (Gond et al., 2011). Consequently, CSR can leverage, for example, recruitment and retention to promote socially more responsible organizational behavior. CSR has the potential to *affect virtually all HRM processes, from pay and compensation to career and talent management, job descriptions or incentive design* (Gond et al., 2011; p. 123). This statement is also emphasized by Fox (2007; p. 44), who stresses that the power of CSR for HR is that it incorporates what HR is already doing, but integrates it with the business and with other key functions of the company and ties the whole package to strategy and the business case.

HRM has capabilities specifically targeted at employees and training, which means that it has the ability to support an organization's CSR performance and value creation. Furthermore, HRM policies and practices provide the framework for an organization's culture and are perfectly positioned to embed CSR values of corporate citizenship into the organizational culture (Lam and Khare, 2010, Inyang et al., 2011). A growing number of researchers are concerned that, by not stepping up to take a proactive lead in sustainable business, HR professionals may be missing out on an important opportunity to capture core competencies that customers will want and competitors will be unable to duplicate (Cohen, 2010; Sroufe et al., 2010; Aagaard, 2012). More recent literatures therefore emphasize that HRM should take ownership of CSR and drive the process holistically from crafting policy to implementing it across the organization while involving employees in the process (Glade, 2008; Melynyte and Ruzevicius, 2008; Inyang et al., 2011; p. 124).

Sroufe et al. (2010: p. 35) underline that being sustainable should include a focus on *developing innovative social and environmental practices that promote collaborative efforts across functions, create unique social capital, and build long-term economic value for a firm.* A similar approach is seen in Ulrich's (1998) suggested role for HRM as "change agents." Generally speaking, CSR chiefly centers on managing changes in philosophy and behavior (Sroufe et al., 2010; Lam and Khare, 2010). Consequently, facilitating a sustainable culture through CSR is viewed as a critical task for HR managers as change agents in integrating CSR and facilitating sustainable business (Sroufe et al., 2010). Through the collaboration with CSR, HRM is provided with the opportunity to contribute to business success, value creation, and improvement of employee engagement (Lam and Khare, 2010). Bhattacharya et al. (2008; 37) therefore stress that CSR *comprises a legitimate, compelling and increasingly important way to attract and retain good employees* and talented CSR-oriented employees that can help foster a CSR-focused culture (Lis, 2012).

6.1 The Link between HRM, CSR, and Sustainable Business

Human resource management (HRM) is basically about how to handle human resources in an organization. In practice, this includes how the company attracts, retains, develops, and lays off employees. HR as a function plays a key role in CSR, as the way in which companies manage, develop, and motivate employees helps to create the profiles that make up the organization, as well as the culture and values that are practices in daily routines. Cohen (2010) therefore suggests that CSR should be anchored in the company's HR in practice.

The role of HRM in relation to CSR and sustainable business has been overlooked in both theory and practice (Wittenberg et al., 2007). This is somewhat paradoxical as sustainable business can hardly be achieved without a transformation of employees' and management's motivation and values (Waldman and Galvin, 2008, p. 333) and without development programs aimed at the training of responsible leaders. These activities are typically handled by the the HRM department, and the successful integration of CSR therefore relies heavily on HRM (Gond et al., 2011). Thus, an overlap exists between HR and CSR, and a clear distinction is not always found between the two areas. Some of these overlaps in assignments relate to, for example, activities in enhancing diversity and equality across the company, establishing internships, and maintaining a healthy working environment and supporting the well-being of workers/employees.

The fact that the HRM function carries out activities with an element of social responsibility also highlights the potential and opportunities of increased cooperation between the two functional areas. The CSR function is often placed in Corporate Communications or in an independent CSR department. However, few companies place CSR as part of the HRM division (Ehnert, 2009) and/or facilitate close collaborations between the two functions. This could potentially be compensated for through the facilitation of joint meetings, activities, and interfaces, where sustainable development areas are discussed and carried out in collaboration.

During the past decade, the emphasis of CSR research has been on the more external side of the organization's corporate sustainability, in relation to sustainable supply chain management, code of conducts, and external stakeholder management. By contrast, the challenges and opportunities of CSR for the companies' internal organization and sustainable management have played a minor role both in research and in the public CSR debate. However, an emerging stream of literature is linking sustainability, CSR, and HRM research (Ehnert, 2009; Cohen, 2010; Pfeffer, 2010). This literature discusses the strategic potentials of sustainability as a concept for HRM to be engaged in. The growing interest in CSR activities targeting, for example, healthy working environments and the focus area of healthy employees has also an impact hereof. In a Gallup poll from 2005, 61% of the respondents replied that these employee-focused HRM activities have a positive financial impact. It is conceivable that the financial performance as well as the level of sustainability of a company is dependent on whether HRM is functioning well in the organization, as "happy workers work better." Another consideration relates to the fact that CSR is basically a holistic approach toward managing an organization. This implies that not only the external organization should function well but also requires that a company has its "own house in order." As such, it does not make any sense to preach accountability and social responsibility to external stakeholders if the company's internal stakeholders (e.g., the employees) are neglected and/or treated poorly. Furthermore, HRM can also play a central role in integrating the CSR strategy across the company and its divisions. At the same time, the HR department is also an important partner for CSR in training managers and employees in new and more sustainable behaviors and skills, for example, code of conducts, green procurement, and waste management (Copenhagen Business School, 2009). In a report from Copenhagen Business School, a number of key areas for collaboration across CSR and HRM are presented (CBS, 2009, p. 15):

- *CSR as part of employee attraction* can help attract new employees and especially young people. In general, more and more job seekers are attaining knowledge about the company's values and social responsibility as a part of their job search. Thus, the firm's CSR profile may assist in attracting new employees.
- *CSR in recruitment* can be incorporated in relation to the company and its CSR strategies on diversity and inclusiveness. HRM can here play a part in incorporating diversity into the practices and objectives of the recruitment process, ensuring that sustainability is linked to the recruitment process.
- *CSR in retention* can be emphasized through a clear CSR profile, which focuses on activities to increase employee loyalty and employee well-being.
- *Employee and corporate reputation* are closely linked. Thus, if the company is recognized for its CSR work and it rubs off on the company's reputation, then it can strengthen job satisfaction among the employees and again attract new employees, customers and partners, which in the end helps to produce positive financial results.

6.2 CSR in Recruitment and Retention

The growth and competitiveness of companies depend heavily on the employees and the knowledge they possess. From the resource-based view, De Saa-Perez and Garcia-Falcon (2002) demonstrates that an appropriate HR system can create and develop organizational capabilities that become sources of competitive advantage. The "internal fit" concept is particularly salient when we examine the organizational processes leading to good firm performance (Wright and McMahan, 1992). There are particularly three sets of HR practices that have been highlighted in the literature in supporting performance:

1. *Training-focused*: An emphasis on skills enhancement and human capital investment
2. *Performance-based reward*: An emphasis on rewarding employees' contributions and outcomes

3. *Team development*: Leadership and team-based activities are extensively developed and carried out.

These three sets of HR practices are critical for developing cross-functional teams and are often interrelated and reinforced by each other (Norrgren and Schaller, 1999; McDonough, 2000). The HRM department plays a critical role in training the right skills and in creating a committed workforce and a positive working environment in which CSR is embedded in all aspects of the work-related life cycle of employees (Weybrecht, 2010). Recruitment and development practices have overlap with the more recent talent management literature, as mentioned by Al-Laham et al. (2011). They stress that in optimizing the benefits from existing human capital, firms must frequently gain access to new knowledge by hiring experienced talent. With the arrival of talent management, recruitment has been elevated to a strategic level, in which innovation plays an increasingly important role. It is therefore necessary for sustainable businesses to link CSR with HR if they want to attract the right people that are able to work diligently and effectively with accountability and social responsibly across the business and the organization in creating sustainable results.

Research has stressed the impact of CSR on prospective employees while making a job choice decision (Turban and Greening, 1997; Albinger and Freeman, 2000; Jones et al, 2010). However, both in practice and in theory, the emphasis has been on how to recruit the younger generation, popularly known as Generation Y, as this generation is entering the labor market these years. This generation is believed to have special characteristics that challenge the traditional ways of recruitment and retention (Tapscott, 2008). According to Tapscott (2008), the Generation Y is characterized by shorter employment periods at each company and higher demands for competence development and more opportunities for career advancement; and if/when their demands for are not met, they quickly move on. This challenges the HRM of many companies and requires new methods of recruitment and retention of employees. One could also question whether it is imperative to retain employees from this generation for years as their skills and competences are developed by moving from one company to the next. In practice, many companies apply existing employees and their networks to recruit similar profiles. In addition, job seekers apply their network's recommendations about a company in their job selection, as other young people's experiences are weighted higher than the company's corporate communications, for example, on their website or public communication materials. A good reputation is therefore created not through the company's formal communications, but through its employees and how the company is portrayed in public.

6.3 CSR as Part of Competence Development

Competence development and training are stressed as key factors in optimizing the fit between the existing and requisite knowledge and skills, while also contributing to performance and knowledge-creation processes in the company (de Winne and Sels, 2010). High-performing organizations tend to spend more time on education and training, especially on communication and team skills (Valle et al., 2000; Leede et al., 2002). Skills development and training are some of many elements that can help to retain employees, because training and development ensure that employees continuously evolve to perform better in existing and new tasks and therefore hopefully want to continue to work for the company. However, one could also stress that companies have a responsibility to keep their employees not only updated and competent for the sake of both the company and their employees but also because of the role they play in society and whose resources they apply to generate positive, business results. Companies have a responsibility to keep their employees "employable"—so that they have the opportunities of getting new jobs, both within the company and externally with other companies (CBS, 2008, 10–11).

The concept of "employability" is therefore one of the factors associating CSR with competence development. Hence, when companies ensure that their employees have the necessary skills, they are also helping to create a more sustainable and competitive society, region, and/or country. This implies in practice that, a company develops employees that will eventually leave one day. At the same time, the company also receives skilled and trained staff from other companies that have invested in training their employees, so the circular economy mindset actually also goes for employees. It is advantageous for companies to send out skilled and trained people into society and the world as it benefits the company's reputation and therefore makes it easier to attract resources, new talents, and partnerships (Cohen, 2010). Several large companies integrate CSR in their skill development programs, as it is now to a greater extent than before a general requirement that employees and managers have the skills and are able to act and work sustainably in a complex global society and across global relationships. An example of such a company and approach is PriceWaterhouseCoopers (PwC) and their Ulysses project.

6.4 Case Example: PwC

The Ulysses program was established in 2001 and has sent over 80 PwC partners from 32 territories on 26 projects in 20 different developing countries.

It was originally designed as an innovative response to the core challenges that many businesses face in an increasingly interconnected global world. The aim of The Ulysses program is to develop leaders, who understand the real-world worth of PwC's values and who can deliver responsible and sustainable business solutions. The Ulysses is also a learning journey that has helped to drive the personal transformation of PwC's leaders. Several leading PwC partners contributed to the development of Project Ulysses, a leadership development and talent management program for top-performing PwC partners, who have demonstrated exceptional leadership potential. Among the many benefits that PwC has gained for the program is the ability to differentiate the organization by the quality of its relationships with clients and the community based on shared values, understanding, and collaboration.

The key objectives of the Ulysses program's are as follows:

- To develop responsible leaders who are capable of assuming senior leadership roles at both local and international levels.
- To build a global network of leaders who understand the importance of values in developing trust-based relationships with a diverse range of stakeholders and who can create a sustainable brand that is differentiated by the quality of their relationships.
- To help PwC leaders understand the changing role of business in influencing the economic, political, social, and environmental well-being of communities and markets across the world.
- To develop a model for Price Waterhouse Coopers that will enable its next generation to lead responsibly within a global networked organization.

The Ulysses program consists of 2 weeks of preparation, 8 weeks on location in a developing country, and a week of review and recommendations. One example of the program's work is the Lokoho Rural Electrification Project in rural Madagascar where four PwC Partners from Russia, France, Indonesia, and the USA participated. The work was carried out in a partnership with the United Nation's Growing Sustainable Business for Poverty Reduction Initiative' and their efforts to guide the selection of projects best suited to reduce poverty and create economic growth in the region. The aim of the project is to remove PwC partners from their comfort zone and challenge them to work in a diversified team in a completely new environment, often in collaboration with NGOs. The program therefore assists in cultivating sustainable leaders by placing them in different work environments and exposing them to complex global problems, for they to learn and to think more responsibly and globally when solving complex problems (Pless et al., 2011).

6.5 Measuring the CSR Effects of HRM

The majority of the HRM literature focuses on HR bundles and claims that it is not beneficial to examine only a single type of HR practice and its effect on a firm's performance. Instead, combinations of HR practices and their contingent effects have to be analyzed (Wright and Boswell, 2002; Bowen and Ostroff, 2004). Sheppeck and Militello (2000) stress that not only different HR configurations are needed to achieve a high level of firm performance but also different types of HR practices generate different outcomes. However, measuring the performance of CSR and the HRM-related activities of a CSR strategy has proven to be a difficult task. Performance in this context represents a broad range of economic, social, and environmental impacts caused by many different business operations, which requires multiple metrics to fully cover the scope and results (Rowley and Berman, 2000; Gond and Crane, 2009). One way of targeting and evaluating effects of the HRM-related CSR activities is to identify and apply key performance indicators (KPIs) targeting the sustainable elements of HRM. Companies can choose to link HRM and CSR activities with the incentive and performance structure of the company and with the managers' and employees' personal KPIs. For example, some companies choose to work with a combination of traditional financial KPIs and specific KPI's related to the employee's or manager's work with company values and corporate sustainability. An case example hereof was illustrated earlier in the case example by Novo Nordisk.

The assessments and KPI's of HRM-related CSR activities can, for example, evaluate the extent to which employees participate in a new CSR policy and the daily CSR activities, or assess the level of employee awareness or knowledge of the company's CSR strategy (Gond et al., 2011). The key success factor of effective CSR measurement is that it has to make sense to those being evaluated and it has to be possible to measure in practice. Furthermore, the evaluation must be motivated in such a way that the employees and managers want to participate and engage if the full impact and results of the company's CSR and sustainable business practice is to be achieved. However, many companies regret of measuring only the scope of CSR activities in relation to their reporting initiatives, but not the business effect of the activities, and that is considered a mistake. For if companies have no idea how CSR contributes to achieving the company's business goals and in making a difference to the company's stakeholders, then it easily becomes a "fairytale" and not an integrated part of corporate business.

Evaluating the effect of sustainability in the HRM function requires both qualitative and quantitative measures to capture the company's ability to gain positive results from its CSR strategy and CSR initiatives across the organization (Aagaard, 2012). Some examples of how to measure the effect of HRM-related CSR activities could constitute the following:

- **Measuring sustainable behavior of employees**
  What to measure: To what extent has the HRM-related CSR activities assisted in creating more sustainable behavior in the internal organization?
  Ways of measuring: Evaluate reductions in the use of resources, raw materials, electricity, gasoline, and so on and the extent of recycling and waste reduction across the divisions.
- **Measuring the awareness and attitude toward CSR**
  What to measure: To what extent has the company's HRM-related CSR activities (e.g., training, seminars, events) assisted in enhancing the level of awareness and knowledge of as well as the employees/managers attitudes toward the company's CSR initiatives?
  Ways of measuring: Evaluate the development in awareness of and attitude toward corporate responsibility and the CSR initiatives carried out among employees and manager through, for example, surveys. This could also be measured across the company's external stakeholders (customers, suppliers, investors etc.).
- **Measuring general working environment**
  What to measure: To what extent has the company's CSR activities (e.g., in management training, conflict management, stop bullying campaigns) assisted in creating a better working environment in the divisions?
  Ways of measuring: evaluate the development in employee satisfaction and management evaluation, as well as the development in and type of conflicts and bullying.

- **Measuring health and well-being of employees and managers**
  What to measure: To what extent has the company's HRM activities (e.g., in retention and training) helped in increasing the physical and psychological health and well-being of our employees and managers?
  Ways of measuring: evaluate the development in the numbers of employees and managers on sick leave, absence, anxiety/stress/depression, number of employees attending sports activities at the company, quitting to smoke, etc.
- **Measuring diversity**
  What to measure: To what extent has the HRM-related CSR activities be able to increase diversity?
  Ways of measuring: Evaluating the (development in the) division of employees based on gender, age, nationality, as well as the increase in numbers of female managers and boards members, number of employees with physical and psychiatric handicaps or with criminal records, and so on.
- **Other types of measurements**
  Many large companies already generate different types of CSR reporting and therefore document the (internal and external) societal and ethical results gained by the company that way. Furthermore, corporate communications typically measure the attention the company receives, for example, by measuring the number of pages in the press. This measurement can be extended to also include an evaluation of positive vs. negative responses to the company and their social responsibility in relations to employee-driven CSR activities, employee health and diversity, and so on. Another measurement approach could be to include health, working environment and diversity as part of the job appraisal interviews that many companies already carry out.

6.6 CSR in Layoffs and Retirements

Layoffs are inevitable. However, the manner in which the employee or manager is dismissed and how this is perceived by the dismissed and the society is critical to the company's brand and reputation and should therefore be handled in a professional and sustainable manner. Social responsiblity plays a key role in layoffs. Yet, this area is not emphasized in literature although central to a company's sustainable practice and reputation (Lakshman et al, 2014). The negative influences that mass layoffs can have on a company's reputation says something about the potential gains of emphasizing sustainability in these types of HRM activities.

In correspondence to this, companies in a growing number have established severance agreements including so-called outplacement packages, where the dismissed employee receives professional supervision and counseling, for example, in preparing a professional CV and in carrying out successful job searching and job interviews. These initiatives testify to a growing understanding of the potential negative effects that a dismissal may have on the company's reputation—both on a micro level (from employee to employee) and on a macro level (in society). Furthermore, the prestige, power, and impact of CEO's are damaged by layoffs. This in addition to the decline in the performance-linked bonuses and loss of firm-specific human capital (Iverson and Zatzick, 2011). An unjust dismissal of an employee can have negative, psychological effects for the individual being laid off, while making it difficult for an employee to move forward. This is of course unpleasant for the individual, but it also has a negative impact on society if that person does not acquire a new job and get back on the labor market. There is therefore a sustainability element in laying off employees in the proper way ensuring that the former employee can continue to contribute positively to society and in another organization (Datta et al., 2010). HRM can therefore play a central role in ensuring that people are laid off in a professional and sustainable manner that makes it easier for the person to move on in another job and in another company.

Voluntary terminations are not documented in the same way in the statistics as dismissals and layoffs. However, there may be many different and unsustainable reasons, why an employee decides to quit. HRM has a role to play in identifying potential unsustainable management or handling of personnel. In practice, they can arrange dismissal interviews with the employees that terminate their jobs to find the possible reasons and to detect potentially unsustainable management practices that HRM has to attend to, while supporting a healthy and positive working environment across the company and its functions and divisions.

Another group of resignations is retirements. Although this kind of dismissal is usually positive, it is still important for HRM to ensure that this is done properly. More and more elderly people request half-time jobs or other job arrangements, where they can stay longer on the labor market, but in ways that take their age and health into account. In continuation of this, HRM can integrate a sustainable senior policy as an element in the company's sustainable HRM and CSR policies.

6.7 Case Example: The Specialists

The Specialists pioneers in several areas. It is internationally known as the first company in the world to use employees with autism as their main labor. The company was originally founded by Thorkil Sonne in 2003 as a private company, when his son of three had just received an autism diagnosis. With the family's house as collateral, Thorkil started the first company in the world providing the proper work conditions for people with autism spectrum disorders (ASD) or popularly known as autism. The Specialists is a social enterprise that aims at changing the world's view of people with autism, while enhancing the level of inclusion for the common good. The company offers IT services and consultancy services to businesses across service areas such as testing, critical IT systems, programming, data conversion, filing, data logistics, and data recording.

The objective was not to create a company that fitted to the owner's son, but to change society's attitude toward people with ASD and to get people to look at autistic people with respect and courtesy and as worthy and valuable members of society. The Specialists is today owned by the Specialist People Foundation (SPF), which is a nonprofit operating foundation, based on support from private foundations and the European commission. Their main office is located in Denmark, but SPF is currently in the process of implementing branches in Poland, Ireland, England, Germany, Austria, Spain, Turkey, and Singapore. In North America, the focus is on the states of Delaware, Minnesota, Maryland, New York, Colorado, California, and Washington, as well as Calgary and Vancouver in Canada.The Specialist People Foundation aims at creating 1 million new jobs worldwide. In September 2009, they established a three-year individually planned secondary education for students with autism spectrum disorders.

SPF is founded on sustainability, and sustainable HRM is a key element in their success as a sustainable business working with people with ASD. However, the typical HRM practices (e.g., talent management) applied in other companies do not necessarily work in this context. Thus, different approaches toward management, organizing, training, and development had to be developed to accommodate the specific needs of employees with autism, as explained by the COO, Henrik Thomsen.

An important element of sustainable management at The Specialist is responsible communication. In practice, this means transcending from child-to-adult communication to adult-to-adult communication. Furthermore, management has had to remove some of the key sources of communication errors, such as irony. *Irony is lost on people with autism, they do not understand it, and it confuses them,* as stressed by Henrik Thomsen. The objective of sustainable management at the Specialist is therefore to ensure that the time that the employees spend working for the company has to be a good experience. At the same time, management must be good at delegating responsibility with freedom and develop a structure and a culture that can handle this. Diversity management at The Specialist is about inclusion and integration. When The Specialists apply the 80/20 rule, it means that management invests 80% to get the extra 20% of the individual employee. *To integrate our employees is something only very few can and to succeed the company's internal and external stakeholders have to be open,* as underlined by COO, Henrik Thomsen.

In developing a sustainable business that could accommodate the leadership necessary for employees with autism, The Specialists had to organize differently than most other organizations. The management level therefore consists of two types of leaders: task managers, who have the technical knowledge, and business managers who are really personnel managers. *In a traditional company you would normally have a project manager who handles both roles. However, when projects and project managers are under pressure, the project manager often runs and pushes the employees harder. Pressure stresses people with autism tremendously and can eventually lead to depression. Therefore, personnel managers are a necessary "regulator valve" in the daily working processes,* says Henrik Thomsen.

Another important element of sustainable management at The Specialists consists of telling the good stories. For example, that Steven Spielberg and Bill Gates, who are both autistic, performed quite well in business anyway. The Specialists have also enhanced their sustainable business by targeting competence development and training to the specific requirements of employees with ASD. One could say, that The Specialists represent a kind of a greenhouse for autistics' talent to unfold. Even though The Specialists have a strong "CSR card" to play to customers, this only works if they provide quality and economic performance. Thus, CSR and business have to go hand in hand to ensure sustainable and financial success over time.

Fundamental to The Specialists and other companies that are built on inclusion and sustainability, is the passion and vision to contribute to a better society. Profit optimization is a premise to continue to exist and do business. Yet, it is not the fundamental value. Some of these sustainable organizations are, as The Specialists, based on a foundation, while other companies are privately owned businesses.

The growth in this type of social businesses like The Specialists bears witness to the fact that a new business type and business model has emerged as a consequence of the global mega trends in sustainable development and the growing interest in "meaningful" business missions. Several companies of this type will shoot up the next several years and will act as an inspiration for new businesses and business concepts, while being instrumental in changing our perception of what a socially responsible company actually is and can be (Ellis, 2010).

6.8 CSR in Administration

The literature and debate on CSR in administration has largely focused on public organizations as the majority of their duties are of an administrative nature. However, private companies and their administrative functions have not at all received the same CSR attention, neither in practice nor in theory. The major sustainability concern of particularly production companies is minimizing their negative environmental impact. Administration does not generate the same level of unsustainable output and maybe that is one reason, why they do not receive the same emphasis in CSR research. Typically, the administration function in private companies is an administrative department working as a direct support function to top management. Due to the organizational position of this function and its close collaboration with management and all company functions, more emphasis should be put on its ability to influence and emphasize, for example, resource consumption across the company and in acting as a role model for the CSR values presented by top management and in the company's CSR strategy. In a report by McKinsey Global Institute (2011), "Meeting the world's energy, materials, food, and water needs," the findings reveal that a key focus for businesses is to administer their resource consumption. McKinsey concludes that resource optimization and productivity will be the single most important competitive factor within the nearest future.

Generally, the responsibility for a company's resource consumption is spread across functions and divisions. However, administration in private companies can play a key role in bridging across the functions in supporting a more sustainable development and in optimizing resource usage and limit waste across the value chain. Heck et al. (2014) reveal that nations as well as companies are already competing for resource productivity and identify

Table 6.1 The actual and required demands for resources in the future

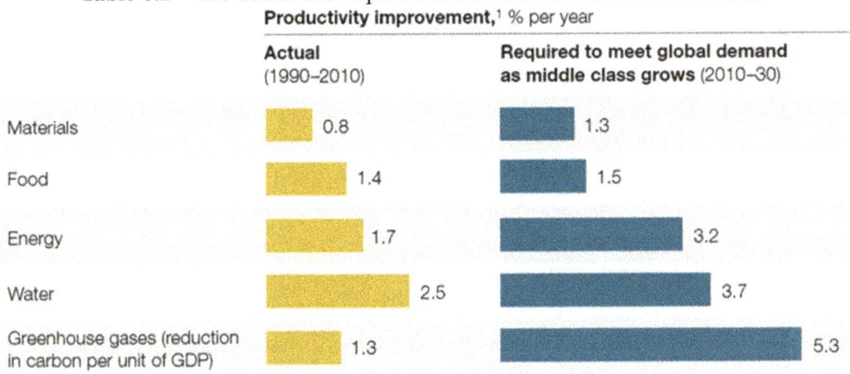

Source: McKinsey Global Institute (2011).

"green" Lean as a key management principle in order to get on the winning team in tomorrow's resource economy.

6.8.1 Lean and Green, Sustainable Administration

The type of tasks that are placed in the administration and the employees who perform these tasks also have different characteristics than the other functions of a company. This implies that the CSR approaches, which have been applied in other functions such as R&D and HR, may not necessarily work effectively in administration. Typical CSR focus areas of administration could be waste management, CO_2 emission, and reductions in water and energy consumption. The integration of lean in administration has been one way that many companies worked with sustainable management in this specific corporate function. However, lean is not CSR. Yet, the two concepts share a number of similarities and can be successfully integrated as both concepts emphasize the necessity to optimize resource consumption, recycling, and waste management.

The link between lean and CSR in administration and service is described by a number of authors and researchers (Eriksen et al., 2005; Tapping and Shuker, 2007; DI, 2007a). In lean, the reduction in waste is stressed for economic reasons, whereas in CSR, the emphasis of lean is on eliminating the environmental footprint. Over recent year, more and more administrative units have integrated lean. This has in many cases been carried out as a follow-up to the companies' integration of other lean projects, e.g. lean production. Yet despite the experience gained from lean production, the integration of lean in corporate administration has not been without challenges. In a study by Danish Industry (DI, 2007a), a number of private-owned companies were asked about their experiences with integrating lean in administration and service. One of the informants, a factory manager from Abena Production, said: *It's harder to get lean to work in administrative units. In manufacturing, employees are more familiar with change, are more flexible and accustomed to sharing tasks. Administrative personnel are more individual and have different tasks. They need to relate to the new framework and uniform procedures, and experience a shift in the power balance.*

If the company has already worked with lean, the mindset of resource optimization can potentially be used to create a better understanding of what characterizes sustainable administration. However, it is important for a successful integration of CSR in administration that management realizes that sustainable administration is about more than printing the annual report on recycled paper. Sustainable administration must be part of the way in which the company chooses to administer and manage its buildings, personnel and payroll expenditures, financing, secretariat, and other administrative services. For all of these activities are images internally and externally on what type of sustainable practices management prioritizes. Furthermore, they all account for the use and waste of valuable resources.

6.8.2 Sustainable Public Administration

Sustainable administration is a concept that has received more attention in public and state-owned organizations. This is really not that surprising as a large part of the public tasks of the state, municipalities and other public organizations consist of, for example, administration of citizens (families, the elderly, children, and young people), housing and construction, jobs and education, health and disease control, nature and environment, and so on, Aall of which generate vast resource allocations and resource use, as well as the potentials for waste and pollution. There is therefore a huge rationale and potential in integrating CSR into the management and conduct of public administration. The shift away from planning that devalues natural capital and increases risk and inequality is another reason for the increasing popularity of sustainability in public administration. Particularly, as one of the primary roles of public agencies is to reduce risk to citizens (Giddens, 2003, p. 34).

More and more municipalities are beginning to incorporate social responsibility into their tasks in a more strategic way. However, only a limited number of municipalities have announced a CSR strategy publicly. Many of the municipalities prioritize climate as one of the main focus areas of the organization's CSR activities. It may initially appear strange, why climate and energy has such a high priority in the public organizations and in government's sustainability emphasis, when compared to so many other important issues. Yet, once you look into the numerous functions that municipalities carry out that affects climate and resource consumption, then it makes perfect sense. As stressed by Fischer (2000, p. 9), modern societies depend heavily on natural capital as a source of wealth, yet struggle to integrate time and place knowledge with scientific knowledge, and both are required for "any effort to develop infrastructure that can be sustained over a long period" (Ostrom et al., 1993, p. 50).

Typical strategic CSR initiatives of public organizations span from green procurement, conversion to sustainable energy sources, sustainable resource consumption, waste management to environmental friendly public transportation—just to name a few. Municipalities affect so many different and elementary things in our daily lives and are therefore important in support the sustainable development and socially responsible attitudes among citizens and in relation to the choices citizens make every day. These choices may for example be choosing between going by bus/train to work instead of by car, or the choice of participating as a volunteer in the local community, attending the school board, and/or to become a visiting friend for a lonely elderly or a foreign refugee. Understanding sustainable development in public organizations also requires a discussion of civil society and opportunities for stakeholder participation. O'Connell (1999, p. 11, 12) defines:

Civil society is a systematic relationship of community, voluntary organizations, government, and business, wherein rights and responsibilities are placed in balance.

Without engaging the civil society in the CSR strategy implementation of the municipal or other public organization, sustainable development can never really become a reality. In practice, strengthening civil society requires recognition of the existence of a problem, renegotiation of the social contract between stakeholders, development of a sense of individual responsibility, investment in social capital, recognition that problems cannot be solved by any single system alone, and shared governance powers, as explained by O'Connell (1999, p. 118).

Although more and more municipalities are dedicating themselves to CSR strategies and sustainable development, they still need to communicate their sustainable initiatives better and more publicly than they do today.

In particular, when talking social and sustainable development that can potentially attract new families and taxpayers to the municipality, for example, energy-efficient housing and efficient and green public transportation. The lack of communication of the sustainable activities of public organizations implies that the full potentials and results of public CSR may not be gained.

Municipalities have to attract "customers" just like companies, not because they have to make a profit, but because citizens help develop and contribute to the growth and development of the society and the municipality. This is also why the municipalities should be more proactive and strategic in their CSR communication and CSR reporting to retain existing citizens and to attract new, potential citizens. Sustainability is an attraction factor for especially younger citizens (Generation Y) and families that typically use more of the municipalities' public services (e.g., kindergartens, schools, sports facilities, hospitals, etc.). However, effective strategies for sustainable development of a society and a municipality rely on stakeholder participation and on understanding the community and the individual motivations (Leuenberger and Wakin, 2007). In the stakeholder management and strategic CSR communication of the municipalities, it is therefore also important to target communication at types of citizens that the municipality wants to attract more of. Thus, many good reasons exits why municipalities and other public bodies should work strategically with sustainability in the administration and communicate it accordingly.

7

Integrating CSR in R&D

Corporate social innovation (CSI), also referred to as sustainable innovation, is characterized by the fact that social responsibility is the focal point of a company's innovation process. Through CSI, sustainability creates new opportunities and new values for the company (Pavelin and Porter, 2008). The companies can thus help to solve some of the world's social and environmental challenges, and at the same time develop new products, markets, and business areas. Furthermore, by working strategically with CSR-driven innovations, companies can increase their growth and competitiveness. This viewpoint is supported by a number of researchers, who have also identified a correlation between company performance and CSI (Hull and Rothenberg, 2008; Wagner, 2010). The theoretical concept of corporate social innovation was originally introduced more than a decade earlier by the researcher, Rosabeth Moss, who identified and explained a new behavior among companies. In her article, *From spare change to real change: The social sector as beta site for business innovation*, she described how companies showed interest in meeting the needs of the social sector and in creating innovation based on these needs. Her point was that companies had begun to view social problems as economic opportunities and that the solutions to these problems were attractive for the social services sector and for companies as well (Kanter, 1999).

In continuation hereof, Nidumolu et al. (2009, 57) argue that *sustainability is a mother lode of organizational and technological innovations that yield both bottom line and top line returns*. They continue by stating that companies succeed in reducing costs by reducing the inputs they use, if they become environment friendly.

The concept of sustainable innovation has emerged due to several reasons such as the pressure from global competition and the growing awareness of the global environmental effect. Although the first stone of corporate production was laid almost three decades earlier, in the 1970s, people began to speak of "acid rain" and "holes in the ozone layer," which addressed the harmful side effects of companies using chemicals and fossil fuels in the production process. The Brundtland Committee concluded in 1983 that companies should pollute less by investing in and by using alternative technologies and materials in their production. Since then there has been a rapid growth in the attention to companies' environmental commitment and it reached the culminating point after Al Gore and IPCC10 in 2007 were awarded the Nobel Peace Prize. From this point, no one could over-rule the problem. Consumers gained much awareness of the severity of the issues and started to demand for sustainable and environment-friendly products, purchasing eco-labeled products communicating the lower environmental and social impact of corporate operations (Loureiro and Lotade, 2005). One could therefore say that several factors have motivated the development of sustainable innovations. Among others, the increased attention in customers, businesses, and the media, combined with the limitations on natural resources and fossil fuels, have increased the focus on alternative forms of energy. In addition, new innovation opportunities have been created through new technologies, and new market opportunities have been emerging in the developing countries.

7.1 Mapping the Concept of Sustainable Innovation

The definition of the concept of sustainable innovation can be traced back decades to the Brundtland report, which emphasizes that *Sustainable development is development that meets the needs of the present without compromising the ability of future generations to meet their needs (Brundtland, 1987).* Thus, sustainable innovation is closely interconnected with the triple bottom line and its sustainable dimensions of people, planet, and profit (Elkinton, 1997). Yet, no single, established definition of sustainable innovation exists, which is why the concept is explained in practice in so many different ways. However, the literature seems to comprise several streams and dimensions defining the concept of sustainable innovation.

The first stream is based on environmental innovations (eco-innovation) described as the development of new services, processes, or products with the ability to decrease the environmental impact (James, 1997). The main driver for business to enter into these sustainable innovations is mostly based on voluntary regulations, standards, and product stewardships (Larson, 2000).

Kaebernick et al. (2003) argued that the introduction of environmental requirements into the product development process at all stages of a product's life leads to a new paradigm of sustainability, which is reflected in a new way of thinking, new applications of tools and methodologies in every single step of product development (p. 468). This argument indicates that integration of environmental requirements into product development must encompass all stages of the products. These life cycle stages of a product are identified as raw materials, manufacturing, distribution, product use, and end of life (Choi et al., 2008).

However, other researchers have emphasized the need to go beyond these compliances and extend the focus to more long-term innovative processes (Boons et al., 2013). In this optic, the concept of eco-innovation is therefore inadequate as it solely addresses the environmental and economic dimensions of sustainable innovation and leaves out the social and ethical dimensions (Charter and Clark, 2007). According to Charter and Clark (2007), the social and ethical dimensions are especially relevant in the case of new, emerging low-income markets where the world's poorest people, the so-called: "bottom of the pyramid" (BOP), are considered as new customers. Consequently, Charter and Clark (2007) offer a new definition embracing all of these elements, which could be considered as the second stream of definitions in sustainable innovation:

Sustainable innovation is a process where sustainability considerations (environmental, social and financial) are integrated into company systems from idea generation and development (R&D) and commercialization. This applies to products, services and technologies, as well as to new business and organizational models (Charter and Clark, 2007: 9).

Additionally, sustainable innovation is perceived as a new business strategy, where social and environmental issues are seen as commercially profitable options and as a source to increase future competitiveness. Another stream defines sustainable innovation by emphasizing the importance of sustainable innovative entrepreneurs with an implicit focus on creativity through sharing, creation, and use of new knowledge in collaboration with different stakeholder groups (de Sousa, 2006; Schaltegger and Wagner, 2011). In this respect, Kanter (1999: 123) defines sustainable innovation as follows:

Learning laboratories where they can stretch their thinking, extend their capabilities, experiment with new technologies, get feedback from early users about product potential, and gain experience with underserved and emerging markets.

Integrating sustainable innovation requires a sustainable foundation. This means that working with CSI is not optimal for organizations until they have operationalized their social responsibility and potentially implemented a CSR strategy emphasizing the sustainability of their business. Thus, if CSR is not anchored into the company's self-understanding, values, and strategy, it will not have the necessary foundation of corporate sustainability to build on. Basically, CSI is about generating innovations and development through sustainability. This implies that whatever strategies companies and managers have learnt beforefor making profit need to be altered and changed for a new and sustainable understanding that emphasizes how profits can be generated through sustainable solutions, choices, and innovations, which ensures positive and sustainable consequences in the short and long run, and across the entire value chain of the company. CSI and sustainable innovation are therefore also a natural Step 2 on the way forward for sustainable businesses in building sustainability into existing innovation and new innovations out of global and national sustainability challenges and needs. Through the companies' learning and experiences with CSR, new business potentials are often discovered and CSI and sustainable innovation are then approaches that can assist the company in leveraging these business opportunities and in generating new venues for sustainable business growth and development.

7.2 Sustainable Innovation Requires Open Source Innovation

Open source innovation is a perquisite for sustainable innovation, as this type of innovation is targeted at society and its stakeholders' needs, challenges and requirements. Chesbrough first coined the term open source innovation in his 2003 book *Open innovation: The new imperative for creating and profiting from technology*. Chesbrough et al. (2006) later defined open innovation (OI) as the use of purposive inflows and outflows of knowledge to accelerate internal innovation and expand the markets for external use of innovation, respectively. Open innovation is a paradigm that assumes that firms can and should use external ideas as well as internal ideas, and internal and external paths to market, as they look to advance their technology (p. vii). Prevalent reviews on the topic emphasize that open innovation entails both exploration of external knowledge and ideas, so-called inbound activities, as well as exploitation of internal knowledge outside organizational boundaries, referred to as outbound activities (Dahlander and Gann, 2010; Bengtsson et al., 2015).

Open innovation is disseminated both through the academic community (Dahlander and Gann, 2010) as well as among practitioners (Van de Vrande et al., 2009). From the present OI research, it appears that companies are pursuing radical innovation through different types of open source innovation activities and across different types of partners (Rass et al., 2013; Chesbrough and Brunswicker, 2014). A plenlty of suggestions for further research in the growing OI literature stress on the research gap of empirical studies in managing and facilitating open source innovation through different innovation approaches, motivating further investigation of effective OI and its different applications (Huizingh, 2011; West and Bogers, 2014; West et al., 2014).

In a CSI report by Danish Business Authority (2009), *Corporate Social Innovation: Companies' participation in solving global challenges*, the CSI tendencies were examined and explained through ten internal case studies from Nokia, Philips, DONG, ISS, Grundfos, Novozymes, Toms Group, IBM, Innocent, and Interface Floor.

The report reveals that one of the important challenges of CSI is the infrastructure and the existing systems of companies, which were formed based on yesterday's way of living, manufacturing, and consuming. Thus, it may be difficult to anchor CSI in the existing structures as CSI possesses a more open approach toward innovation. CSI therefore requires that companies restructure the way they create business innovations, as well as the way they work with customers and stakeholders throughout their open source innovation.

The opposite of open innovation is closed innovation. Originally, Henry Chesbrough, who is an adjunct professor and executive director of the Center for Open Innovation at the Haas School of Business at the University of California, Berkeley, invented the concepts. He explains the basic difference between the closed and open innovation model is that in an open model, know-how and technology can pass in and out of the company's product over time. In a closed innovation model, developments are being made within the company framework without involvement of clients and external partners (Chesbrough, 2005).

Over the past decade, the boundaries between a company and its customers, partners, and the environment have been reduced, among others due to social media and other technologies that support sharing of ideas and knowledge. Thus, customers, suppliers, and other relevant stakeholders should be included more actively in the company's innovation to ensure that their ideas as well as their knowledge, demands, and requirements are used and integrated much more effectively in creating new and sustainable innovations. The basic idea behind open source innovation is that in a knowledgeable society and in a global world of knowledge, companies cannot afford to solely build their innovation on the company's own research and internal knowledge resources. Companies should therefore put more effort to obtain licenses on processes, product ideas, and patents from other companies and stakeholders.

Similarly, internal inventions that are not used in a business should be sold or transferred to other companies, for example, through licensing, joint ventures, or spin-off so that the restrictions that the company framework sets do not stop innovation (Aagaard, 2011).

A number of researchers stress on the disruptive circumstances that through external stakeholder pressures often lead to the creation of radical, sustainable innovations while sustaining circumstances, where, for example, customers are willing to accept minor product adjustment, typically lead to incremental sustainable innovations (Christensen, 1997; Steketee; 2010). Researchers indicate that companies have recently moved beyond eco-efficiency compliances and extended their focus to the adaption of disruptive innovative processes where businesses respond with new game-changing business models (Schaltegger and Wagner, 2008; Loorbach et al., 2009; Boons et al., 2013). This evidence emphasizes the potentials of companies in pursuing both incremental innovations through the perspective of eco-efficiency in products and processes and in pursuing more radical innovations through business model innovations.

7.2.1 Sustainable Innovation in SMEs

In particular, small- and medium-sized enterprises (SMEs) have an advantage when it comes to sustainable and CSR-driven innovation. This was the conclusion drawn from a study prepared by Harvard Business School in 2005. The study examined the competitiveness and growth potential among SMEs, and 66% of the interviewed SMEs found that CSI contributes to a positive economic result. However, despite this, only 26% of the interviewed SMEs developed CSR-oriented products or services, indicating that this is a huge development potential (Harvard University, 2005).

The study also underlined CSI as the most significant competitive tool for SMEs. This is partly because of SME's flexibility, which enables them to adapt their production and processes to new emerging niche markets. They can therefore quickly tailor and differentiate their products from their competitors. Later, Bos-Brouwers (2010) conducted an empirical study on sustainable innovation among SMEs in the rubber and plastics industries. The findings showed that many of the sustainable innovations were aimed at improving the technological processes (eco-efficiency) and introducing lower production costs. By contrast, the companies that had sustainability integrated into their vision, strategy, and innovation processes created value by developing sustainable products that were new to the market, and by working with stakeholders.

One practical example of a sustainable and CSR-driven innovation is the product, Life Straw, which is a water purification tool containing a textile micro-filter that purifies water. The product is designed as a large plastic 'straw' and is developed by the Swiss-based company Vestergaard Frandsen. The filter in the suction pipe allows people to drink impure water directly through the suction tube through which the water is purified. The invention is especially intended for people in areas with poor water quality, typically in third world countries and in areas affected by natural disasters. Other international examples of CSI include the following:

- *Nokia Data Collection*, which consists of a software that enables the exchange of questionnaires and interview results via the mobile network, particularly useful for NGOs such as the Red Cross and UNICEF when sharing data and work projects across borders and great geographical distances
- *Grundfos Life Link*, which has developed a water pump powered by solar cells
- *Philips' Smile Project*, which has developed diode flashlights used where the consumer does not have access to electricity or batteries

The inspiration for sustainable innovation often comes from outside, and either through the requirements of the customers and/or in relations to sustainability issues in society. Oftentimes, companies lack the right idea and approach of how to pursue these market opportunities as they do not have any guidelines, tools, systems, or processes for CSI installed in the organization. The lack of knowledge and inspiration are often the key barriers holding back the company's work with sustainable innovation. Furthermore, the unique characteristics, values, culture, history, and existing product portfolio of the individual organization can also affect the way in which CSI is incorporated. Thus, all of these elements should be incorporated and considered when developing the company's own CSI strategy. Organizations must of course be careful not to blindly adopt other companies' CSI approach, but instead collect knowledge about opportunities and requirements for CSI among their own internal and external stakeholders, partnerships, and the communities that they are associated with.

7.3 The Customer in the Center of CSI

CSI does not replace CSR, but should be considered as a supplement and complementary to the other CSR activities that the company is already carrying out. Consequently, CSI can assist in anchoring sustainability into the company's innovation and R&D department.

Profit is not the sole or primary aim of a CSI-based development project, yet it is still an objective as CSI is part of a sustainable business and not about philanthropy. Basically, CSI is about creating sustainable business opportunities and generating new products that can help improve specific, social, ethical, and/or environmental challenges that stakeholders are experiencing in a community and/or in the world. Thus, the idea process of sustainable innovation begins somewhere other than traditional innovation and R&D, where the technical capabilities often set the limits. In CSI, society and customers represent critical information sources in identifying the unsolved local, national, and global challenges and the elements of sustainability that are most important in relation to the product and service types and features that the company develops and provides (Ellis, 2010).

CSI therefore requires an open source innovation approach, which in practice implies that the company invites customers, partners, and stakeholders into the innovation process to identify and integrate their requirements and needs into the development of customer-oriented and sustainable products and services. The traditionally closed innovation approach toward R&D customers and partners are only involved to market the products where engineers conceive what they think the market wants. As both emphasized in theory and practice this approach is outdated, since the customer's power and access to knowledge has increased rapidly over the past decade, which makes them key players to be included in the development process to ensure product market success (Chesbrough, 2005).

However, what customers are saying and what they do are two different things. A study by the Danish Chamber of Commerce showed that 92% of respondents believe that consumers should be accountable to the community. The respondents also believed in principle that they have a responsibility to the community, when they as consumers shop, whereas 48% of the informants stated that they as consumers to a large extent should show responsibility, for example, by choosing organic produced, fair trade products, meat and eggs from uncaged hens, and low-energy electric light bulbs. However, when evaluating how responsible the customers are in relation to their actual actions and choices, the picture changes. Only three percent said they always choose sustainable products, while just over half of the informants answer that they often choose the sustainable/responsible products. Although almost half of the consumers report that they greatly should take a social responsibility through their shopping, and more than half say they often choose responsibly the actual sales' figures for total sales of sustainable/responsible products and ecology and fair trade reveal that only seven percent of the grocery trade is made up of these products (Danish Chamber of Commerce, 2008a).

The concept of "cradle to cradle," as explained earlier under the chapter 'Integrating CSR into Production and Procurement,' is central in the mindset of CSI. A CSI approach is to solve or eliminate a sustainability issue through the unique features of a company's sustainable products and/or services. Another approach to CSI is to create new innovations with no or lower environmental impact.

For companies that already work effectively with open source innovation, the transition is not as complicated. However, companies with more closed R&D departments, and where customer dialogue and cooperation are not as well developed, face a major development task. CSI is rarely successful unless it involves the company's customers, partners, and stakeholders, who influence the company's work and development (von Hippel, 2005).

Another critical task in CSI is to communicate and sell the sustainable product in a way that also ensures business success. This often requires training of both marketing and sales representatives in emphasizing other benefits than price and underlining the sustainable business benefits that the product provides the customers, other stakeholders and society. The public debate on CSI has at times pointed toward a perception of the concepts that could resemble pure philanthropy. However, this is a utopian and a misguided approach because companies cannot survive without profits. It is therefore conceived as perfectly all right for companies to generate good business results as long as companies generate and apply the profits in a sustainable way.

Key questions to ask for a company that considers the CSI approach could therefore constitute some of the following considerations (Aagaard, 2012):

- How sustainable is the company's existing production, products, and services?
- What alterations does the company have to make, to develop our existing products and services into more sustainable variations?
- What elements of sustainability are of high priority to the company's customers, stakeholders and partners?
- In what way can the company incorporate these elements into the existing and in the development of new products?
- Who should the company partner up with to be able to create and market new sustainable innovations?
- In what ways can the company develop its existing business and new business models that can support new, sustainable product innovation?

These questions can assist the company in mapping the critical considerations to be made in their pursuit of sustainable innovation. However, the answers require stakeholder dialogues and partnerships with customers, users, suppliers, NGOs, and other stakeholders in ensuring successful integration.

7.4 Case Example: Henkel

Henkel is organized into three globally operating business units: Laundry & Home Care, Beauty Care, and Adhesive Technologies. The Dax-30 company is headquartered in Düsseldorf, Germany, and employs 50,000 employees from more than 120 nations worldwide. The company follows a long tradition of developing more sustainable, innovative products. As an example, they launched the first solvent-free glue stick in 1969 and phosphate-free laundry detergent (Le Chat in France, Persil in Germany) in 1986, which was 10 years before legislation required such a change.

Acting responsibly has been a part of Henkel's DNA ever since the company was founded in 1876.

In the 1950s, Henkel replaced natural soap with synthetic surfactants. These unfortunately foamed not only in the machine but also in the discharged waste water, which led to regular ecological quality checks of all laundry detergents in 1959 as well as the development of surfactants that are biodegradable. In 1969, the company launched the enzyme-containing Persil 70. It was given the claim "biologically active." The pioneering work in enzyme research was the basis for providing superior cleaning performance at lower temperatures. Since then, electricity consumption per machine load has been halved due to this research.

In the 1970s, the company CEO Konrad Henkel declared that *the times are over, when an entrepreneur can concentrate only on maximizing profits and on the health of the company.*

Launching the first phosphate-free Persil in 1986, Henkel responded to both new ecological trends and consumer expectations. Consumers experienced themselves that the new phosphate-free formulation did not compromise cleaning performance. Henkel later started to develop more compact detergents. That way, R&D could reduce the amount of raw materials and packaging as well as the transportation impacts. Whereas 280 grams of conventional washing powder were needed for one wash cycle in the 1970s, just 67.5 grams of Persil or Le Chat Megaperls is now sufficient. In the 1990s, Henkel launched "Persil with Plantaren," containing surfactants based on

100% renewable raw materials. However, that proved to be too abstract for the mass market and was not successful in the market. During the years, the company learned that a significant proportion of consumers are not willing to accept compromises in terms of product performance and convenience.

Over the past 10 years, the company has developed a hypoallergenic range of products, launched new formulas that perform better at low temperatures like 20° or 30° to reduce the energy footprint during use, and developed a pre-dosed detergent to help consumers avoid overdosing and a perfume- and colorant-free formula. In 2013, Henkel unveiled an eco-certified range (Le Chat Eco EcoEfficacité), the first mass market brand to be certified, and in 2014, they began making the refill bottle out of 100% recycled plastic. In 2015, they launched the first auto-dosing bottle that helps consumers to avoid overdosing liquid detergent.

In terms of reporting publicly on sustainability impacts, only a handful of other companies have been reporting as long as Henkel, as the company published their first annual Environment Report in 1992. After this, the company has published over 25 annual CSR report. Meanwhile, sustainability has become more and more important for the consumers and retailers of Henkel. For some of the company's brands like Le Chat, a subrange with eco-label certification has been developed, which represents a strong business opportunity for Henkel. However, as stressed by General Manager of Henkel Laundry & Home Care France, Jean-Baptiste Santoul: *Sustainable innovation requires the full engagement of all teams throughout the entire value chain, from R&D and production teams to the marketing and sales departments.*

Henkel's head of sustainability management, Uwe Bergman, states that *it is worth mentioning that for many of our products the greatest environmental impact occurs during the use phase. So we also concentrate on developing products that enable the efficient use of resources such as energy and water.* Furthermore, Henkel strives to educate consumers and to promote a responsible attitude toward the use of their products through targeted communication. As a consequence, a web campaign called "Lavons Mieux" (wash better) was launched in France 2014.

Successful innovations have to combine outstanding product performance with responsibility for people and the planet—and thus contribute to sustainable development and Henkel's economic success. Sustainability and economic business are therefore essential elements of sustainable business and innovation at Henkel, and as stressed by Jean-Baptiste Santoul, *sustainability goes hand in hand with business performance. It strengthens and secures the long term value of our brands,* which again is supported by Henkel corporate values, which are to deliver more values for customers and consumers at a reduced environmental footprint.

The benefits of sustainable business and innovation for Henkel are among others that sustainability is an important competitive factor in the marketplace, as emphasized by Uwe Bergman. The second benefit is reputational, which has also contributed to Henkel being selected as a preferred employer and partner. Also, Le Chat, Henkel's main detergent brand is considered by retailers in France as the number one detergent supplier in terms of sustainability. And there is also an internal benefit: As it clearly contributes to teams' engagement, as emphasized by Jean-Baptiste Santoul. Especially on Le chat, as the sustainable marketing strategy has been awarded several time as best practices. The biggest challenge in working with sustainable business and sustainable innovation relates to *changing habits and mindsets*, as stated by Uwe Bergman. Furthermore, there are different cultural and market contexts. Consumers in Germany are somewhat more concerned with energy and carbon emissions than consumers in France due to differences in the energy mix and regulatory environment. On the other hand, French consumers tend to rely more on independent eco-labeling (such as the European eco-label) than in Germany. These differences make sustainable innovation a more complex topic.

Sustainable innovation requires changes to strategy, management, HRM, communications, and R&D. In order to implement our strategy and reach our sustainability goals, Henkel has defined three strategic principles: Products, partners, and people. The products are where Henkel can make the biggest difference and this is why they are strongly focused on creating more sustainable value for customers and consumers through innovative solutions and education. The second strategic principle is partners, and by standardizing supplier assessments and sharing the results among member companies, Henkel is reducing costs for member companies and our shared suppliers.

The third strategic focus area is people. Henkel offers their employees a wide range of training and educational programs focusing on sustainability, including the sustainability ambassador program, which they launched in 2012, with the aim to help the staff to gain a detailed understanding of sustainability and to convey this knowledge competently to their co-workers, suppliers, customers, and consumers as well as to students. More than

3,800 employees have already qualified as sustainability ambassadors by successfully completing an online training on the principles of sustainable development and on Henkel's sustainability strategy.

Sustainability can only become an integral part of people's daily work if all employees understand the underlying principles and have the opportunity to make their own contributions. This is why we introduced Action Plan meetings at the end of 2011, when we launched the Sustainability Strategy 2030. During these meetings, managers from all levels of the hierarchy and their teams developed a sustainability action plan for their own particular areas, by defining both short- and long-term measures for achieving the sustainability targets.

On a corporate level, Henkel systematically anchored the focal areas of their sustainability strategy in their corporate innovation process in 2008. As a result, their R&D department and researchers must demonstrate the specific advantages of their project in terms of delivering more value and reducing the environmental footprint. They have also developed the Henkel Sustainability Master, an evaluation system to help researchers assess new innovations. It includes a matrix with the individual steps of our value chains on one axis and our six focal areas on the other axis. The matrix can be used to compare the sustainability profile of two products or processes and assess the changes. In addition, Henkel opened the Henkel Innovation Campus for Advanced Sustainable Technologies (HICAST) in partnership with RWTH Aachen University in 2014 so that their Laundry & Home Care experts could work together with academic experts to develop new sustainable products and technologies. Also in 2014, Henkel partnered with an online crowd-sourcing platform and invited the creative community to submit engaging films showing how consumers can shower more sustainably. Furthermore, their marketing team has undergone special trainings on sustainable marketing and communication and work closely with their corporate sustainability and corporate communication departments, as well as external sustainability consultants.

A key conclusion is that sustainable innovations at Henkel are highly dependent of partnerships. *It is like an ecosystem, it's about the way you interact with stakeholders and communities: Sharing, making sure they leverage the full power of your innovation, listening to their needs, feedback and suggestions. What's the benefit of laundry brands that can deliver full performance at 20° if everyone washes at 60°?*, as underlined by Jean-Baptiste Santoul.

8

Integrating CSR in Different Industries

The CSR literature has so far paid disproportionate attention to CSR integration in larger organizations (Spencer and Rutherfoord, 2003) and has primarily focused on manufacturing/production companies (Williamson et al., 2006). The fact that the manufacturing industry has been popular in CSR studies is quite understandable as companies in this industry through the consumption of natural raw materials and production of physical products generally consume more resources and pollute more than, say, service companies (e.g., banks and insurance companies). However, in recent years, more and more service companies and public organizations have decided to work strategically with sustainable business, and the question is whether companies in these industries should use the same instructions for CSR as production companies.

Both researchers and practitioners discuss the necessity of looking at CSR in relation to the type of business and the industry that the company is part of. The reason being that there is a huge difference between carrying out sustainable business in a manufacturing company and in a service company or in a large multinational company compared to a SME. Neergaard and Pedersen (2007, pp. 90–91) emphasize the need to look at what type of CSR activities that works best in which contexts. Practical experience shows that CSR should be tailored to the context in which it is carried out. Otherwise, it is not possible to achieve an optimal coupling of sustainability and business, and CSR strategies and initiatives are rarely rooted effectively in the organization. Furthermore, there are differences in how far the various industries have reached in getting CSR integrated and embedded in their overall business strategies.

In evaluation of CSR integration, the UN Global Compact Management Model can be applied. The model consists of six focus areas:

1. Management and organization
2. Opportunities and risks
3. Strategy and policies

4. Implementation
5. Measuring and monitoring
6. Communications

Each of the six focus areas are central in implementing sustainable business and anchoring CSR into the company's overall business strategy. Through these six focus areas, a report by Deloitte evaluated the different industries' implementation of CSR. The report concluded that health care takes a clear first place as this industry scores high on all the six focus areas. Second, we find the production companies. At the opposite end are consultancies and service companies that perform significantly lower than average. In the next section, practical CSR implementation cases are presented, respectively, from a manufacturing company, a service company, a public organization, and a SME to illustrate some of the differences in the ways that sustainable business is supported and how CSR is successfully integrated in practice across different company types and industries.

8.1 CSR Integration in Manufacturing Companies

Sustainable business has long been practiced in manufacturing companies. Concepts such as sustainable production and green procurement, responsible supply chain management, recycling, and waste management are well established in practice in production companies and well described in research (Ranganathan, 1998; Carter and Jennings, 2004; Ciliberti et al., 2008) as also discussed in earlier chapters. The public CSR interest in this particular type of companies is largely due to their significant and visible impact on the environment. Manufacturing companies play a significant role in relation to the environment, the global society, and the world economy as they are responsible for comprehensive resource consumption. Furthermore, the manufacturing companies engage in multiple national and global supplier partnerships, which are required to produce, sell, and transport physical goods. Thus, the manufacturing companies affect the social conditions in other societies and countries as well through their supply chain (Sarkis, 2001). In particular, the environmental considerations have from the start had comprehensive attention in the CSR strategies in most manufacturing companies. The world is experiencing a global resource shortage and due to that growing prices, which forces manufacturing companies to prioritize optimization of resource use, if they want to stay competitive. Furthermore, the requirements of governments and global societies stress manufacturing companies to minimize their negative environmental impacts and "footprint" through more sustainable production (Gupta, 1995). Similarly, the suppliers of the manufacturing industry constitute an important part of the production companies' value chain, which has led to an increased emphasis on ethical and sustainable supplier partnerships. The massive growth in companies working with suppliers outside their home country has created a lot of legal "gray areas," which have previously been utilized and/or exploited by producing companies across the world. However, social media has through the years provided access to knowledge of these "shady" business approaches and several cases have been revealed in the press addressing everything from child labor and "slave-like" working conditions, as well as the use of toxic and dangerous chemicals and ingredients in production processes and products. One example hereof are the so-called "sweatshops", which are large clothing factories, typically based in South and Central America and Asia, where employees and children work countless hours without breaks and fair pay and without opportunities for entering into trade unions, and so on. However, with time and the public display through press and social media, public focus has been shifted on these unsustainable practices. Several major international organizations have launched various activities to create awareness of and common set of rules for workers rights worldwide. For example, International Labor Organization, an United Nations initiative, deals with labor issues related to international labor standards and the Clean Clothes Campaign (CCC), which is the clothing industry's largest global alliance of trade unions and non-governmental organizations.

The company's geographical location and decisions on expansions and relocations of production facilities are heavily influenced by international, environmental laws and the environmental approval processes in the specific countries. In many Western countries, the production facilities of companies are by national laws only allowed waste up to a certain level (in the form of pollution of air, water, soil) and if they go beyond that point, they can be fined. Thus, if the company cannot keep within the limits, they must either convert to environmental friendly production or move their facilities to a geographical location where environmental instructions and legislation are more lenient. For example, the multinational company Proctor & Gamble's decisions to expand their facility in Oxnard, California, was almost blocked, because of their inability to meet the permitted requirements for air pollution, and many similar examples have been witnessed since (Dretler, 1997). The solution is for many of these companies to move the less environmentally friendly production to developing countries, where labor is cheap and disorganized, and where the legislation on environmental waste and pollution is minimal or not enforced. The extensive CSR focus on sustainable supply chain management and code of conducts is therefore born out of and as a consequence of previous, non-sustainable behaviors and practices. Another key CSR focus area for manufacturing companies relates to work-related injuries and work safety among production staff. Thus, more and more companies have developed policies for employee health and safety as part of their sustainable supply chain management and code of conducts.

All of these unique characteristics and parameters of manufacturing companies have resulted in the development of a specific CSR approach and emphasis in production companies' sustainable business where CSR is incorporated both indirectly and directly through environmental and quality policies, employee and security policies, and supplier policies. As a consequence, CSR and sustainable business is particularly visible among manufacturing companies, and where several CSR flagships, for example, Adidas, BMW, Siemens, LEGO, and Novo Nordisk have influenced the debate and the development of sustainable business during the past two decades.

One production company that for years has operationalized CSR and sustainable principles across its organization and value chain is LEGO. In the following case, some of the key initiatives in CSR integration and facilitation of sustainable business across LEGO are addressed.

8.1.1 Case Example: LEGO

It is a fundamental value of the LEGO group to "behave properly," so it is therefore difficult to set a time for when LEGO originally started working with CSR in the way we know the concept today. However, one thing is clear, the initial corporate CSR initiatives started emphasizing the environment and responsible supply chain management. In 2006, LEGO decided to arrange all its CSR initiatives under one umbrella, when they saw a need to strengthen their strategic deliberations. LEGO has a clear understanding of CSR integration that it is only possible to work seriously with sustainable business if it is integral to the way business is done. That means that CSR must be implemented as a natural part of the company's business processes. For the same reason, many of the CSR initiatives and focus areas have been integrated and carried out in a decentralized manner as explained by CSR Vice President, Helle Kaspersen. LEGO wants to build their sustainable business in ways that ensure a sustainable basis for the benefit of future generations. This is due to the fact that LEGO's core focus is children, who they also consider their key stakeholder. The strategic guidelines for CSR in

LEGO globally are developed in Denmark, where LEGO is located. However, this does not imply that the way CSR in done in Denmark has to be done the same way in all the other countries where LEGO's subsidiaries are located. For example, on the US market, the CSR focus has long been "doing-good" in the community while emphasizing donations and philanthropy. Yet, over the recent years the US market has emphasized more focus on the environment.

LEGO works from an impact approach, which means that their CSR emphasis is based on the areas in which the organization can make the biggest sustainable difference. *It is important that all employees know what we as a company prioritize in the CSR field and the direction and the goals we have set. CSR must be implemented in the company in a way, so that employees feel responsible for this agenda and implement it in their daily work without each initiative necessarily having to be aligned,* says CSR Vice President, Helle Kaspersen.

LEGO has implemented a number of different CSR activities in the employee area and work through the health and safety standards OHSAS 18001 in all LEGO locations with more than 100 employees, and has among other targeted reducing absenteeism and workplace accidents, and so on. The targets are established following a top-down and bottom-up process, where the overall CSR goals of the company are communicated to each location, which then assess and provide suggestions for the results they expect to reach, and which initiatives they have planned to achieve these objectives.

In procurement, the following question is asked, when making the most sustainable procurement choices: What is the "impact" of the equipment? For each investment LEGO makes, an evaluation of the "impact" on both the social and environmental side is carried out in order to ensure that the investment meets the company's overall objectives in the field. In production, LEGO emphasizes energy efficiency and also focuses on waste management with targets stressing that at least 85% of the company's waste must be recycled, which puts great demand on waste management and sorting waste. Furthermore, in R&D, the LEGO group applies a sustainable product design tool called, "Design4Planet", which focuses on minimizing the environmental and climate impact of LEGO products and production across the company's global value chain. In addition, the CSR department works exclusively with responsible supply chain management at the strategic level and in close cooperation with the operational level through the procurement function. All supplier partnerships are subject to code of conducts and requirement specifications for product quality and safety.

The way LEGO has anchored sustainable business is by setting broad goals, which in turn are divided into business areas and others into processes. LEGO has a wide range of non-financial targets that management follows up on each month and report via a balanced scorecard, which is sent out quarterly to all the business areas. The board of directors and top managers receive this report, and it is available on the LEGO internal website for all employees to read. Likewise, there is a monthly follow-up and articles are published from the various business areas where stories of sustainable progress are presented.

In the prioritization of CSR focus, Helle Kaspersen stresses that *businesses should start thinking about what it is that is important for their business. For not all CSR initiatives make sense for individual companies. Therefore, a company must be careful not to look at all sorts of successful CSR organizations from other industries, which have a different history, background or vision . . . And make sure that the vision for CSR is integrated into the company's vision – and that it is anchored at top management level.*

Likewise, priority is important. The area is so large and there are so many points that can be worked with, all of which can make sense, but which can also help to provide a very diffuse understanding of the agenda and worst case end up with nothing happening. Focus on specific, concrete objectives and areas that the company will prioritize. This makes it easier to work with and easier to communicate – internally and externally.

The manufacturing industry has because of the early demands for environmentally conscious production and due to a logical rationale as resource consumer been forced to integrate CSR into their business. Despite a longer learning curve with CSR than most other industries, the production industry is not finished with embedding sustainable business. Many large manufacturing companies still face the challenge of anchoring the parent company's CSR strategy into their foreign subsidiaries, which have other environmental laws, union requirements, and other beliefs and understandings of sustainability. Enterprises should determine the extent to which their parent companies should interfere and set limits and boundaries for the subsidiaries' CSR activity. At the same time, they must also ensure that the company as a whole still has a common set of values and a common basic understanding of sustainability that can be communicated toward the company's many different and global stakeholders.

8.2 CSR Integration in Service Companies

In the earlier mentioned report by Deloitte (2011), the findings revealed that service is performing significantly below average, when it comes to CSR integration across the organization. A survey of users of mobile telecommunication services conducted by Hongwei and Yan (2011) states that both CSR and service quality have a direct impact on customer satisfaction and identification with the brand. Ergo service companies that have a high quality of service can improve the identification that the customer has with the company brand, by also working with and communicating in a strategic fashion about their corporate responsibility. Results from this study emphasize that CSR can leverage the company's brand and customer identification with the brand.

Increasingly, CSR is viewed as an integrated practice in service companies although it is still manufacturing companies that have the lead in terms of visibility and media coverage of their sustainable practices. One reason is probably that service organizations have not been scrutinized and measured to the same extent in relation to environmental impact and energy consumption as production companies. Public awareness and social media coverage have contributed to the fact that production companies early on felt the pressure to assume their responsibilities toward society and their internal and external stakeholders. Only within recent years has this been requested of service companies, among other due to initiatives like Global Compact and changes in national accounting laws of listed companies' responsibility to report via the triple bottom line.

Typically, the environmental impact from production companies is greater than from service companies, which supports a more intense public focus on CSR in production. However, services are often part of the production of a product and/or lead to production among its subcontractors, so the environmental and social impacts of service companies are equally important. By the mid-1990s services accounted for almost two-thirds of world GDP, up from about half in the 1980s, and these numbers are still growing[1]. This also implies that the service industry employs a large percentage of the global workforce and therefore has an impact on the social development of nations.

Many service companies initially had to design and develop their own CSR approach, as little empirical knowledge of CSR integration was

[1] www.worldbank.org

targeted service companies, and because service companies could not directly adapt the CSR approaches and integration models that production companies applied. However, CSR integration in service companies has gained both theoretical and empirical attention in recent years. Today, more and more service companies publish CSR report applying corporate sustainability in a much more strategic and targeted fashion in their corporate communication and branding. The contents of CSR strategies and CSR report in the respective manufacturing companies and service companies are also starting to resemble each other more and more. The reason for this may relate to the fact that global companies today seek inspiration among the top CSR performances no matter what industry they originate from. Yet, this standardization or "one size fits all" approach to CSR must be avoided, as service and production companies have very different framework conditions, which affect the key priorities of the CSR strategies in the respective industrial contexts. The next case example of CSR integration is with RSA and Codan, and illustrates how a global service company can integrate CSR into the company's business and in one of its subsidiaries.

8.2.1 Case Example: RSA and Codan

The Codan corporation is Scandinavia's third largest insurance company with the flagships Codan in Denmark, Trygg-Hansa in Sweden, and Codan also operates insurance business in Norway. By 2008, the Codan group was acquired by the British company RSA, which is among the largest international life insurance companies with 13.500 employees and more than 20 million customers worldwide spread across 130 different countries. The company's primary products are insurance products to both corporate and retail customers as well as health insurance. RSA has a mission to be "the number one sustainable insurer" in the world. It places great demands on sustainability efforts throughout the group—from the implementation of sustainability in the business and to reporting CSR. Based on RSA's CSR strategy, Codan has implemented four CSR policies, which include the following:

1. General CSR policy (general principles of our behavior as a company)
2. Policy for human rights
3. Policy for environment and climate
4. Policy for donations

RSA Scandinavia has approximately 3900 employees in the Nordic region. In 2010, Codan and Trygg-Hansa received 776,240 claims in Scandinavia, which corresponds to 2130 claims per day, and where the Scandinavian division of RSA paid about 2 billion dollars in damages.

RSA has a long tradition of working with sustainability and CSR, and focus on climate in businesses is almost a prerequisite for doing business in the UK. The UK applies many different CSR rankings that a company can pursue and be measured according to.

In 2007, the subsidiary, Codan, developed a CSR strategy based on input from their key stakeholders (customers, employees, partners, etc.) and based on RSA's strategic CSR focus areas.

As an insurer, it was natural for us to introduce CSR in the organization. We are the financial safety net that helps individuals and firms when the damage is done. With our knowledge and data, we can also help to prevent injuries and accidents—for the benefit of our customers, the community, and our business. We believe fundamentally that companies achieve the greatest success if they manage to combine economic performance with regard to people and the environment", says Codan's CSR director, Jeanette Fangel Løgstrup.

British companies typically emphasize the community aspect and corporate volunteering as it fits their model of society. This approach is, however, not as pronounced in Scandinavia as companies here enjoy the benefits of Scandinavian countries' efficient welfare systems. However, as part of their British allegiance, Codan selected to pursue the idea of corporate volunteering and has entered into a strategic partnership with WWF around a common activity called 'Earth Hour', which is about climate change and emphasizes that all people on the planet turn off the lights for one hour. *I was even out with 60 employees from Codan to promote Earth Hour to local shops in Copenhagen. The employees were almost "high" after such a day!* emphasizes CSR director, Jeanette Fangel Løgstrup. In addition, Codan has introduced an annual volunteer week. In Denmark, they, for example, passed out reflectors in favor of The Children's Accident Foundation[2] and worked as volunteers for the NGO, The Night Owls. In Sweden, the employees worked as volunteers for the Red Cross, WWF, and the Swedish aid organization, Myrorna. Based on Codan's CSR strategy and positive experiences with corporate volunteering, they have decided to

[2]Børneulykkefonden.

give all employees the opportunity to spend 2 days a year on volunteering with pay.

Climate is a key CSR focus area of Codan. In 2010, the insurance industry paid out about 200 million dollars just in Denmark for weather-related damages, including water damage. In 2011, it is estimated that the cost (just in Denmark) will increase to approximately 500 million dollars and much more when you incorporate the rest of RSA. This is an increase of about 50% from 2006 to 2010. Consequently, climate and climate change are some of Codan's key strategic themes of corporate sustainability. A main focus of RSA and Codan's climate actions is to influence citizens to adopt a more climate-friendly behavior. In early 2011, Codan launched a new solution with a 24/7 window service. Hence, customers with glass damages would have new and energy-efficient windows installed within 24 hours. Additionally, glaziers offer customers a free window check of the remaining windows in their houses, and they prepare an energy report that shows how much energy and CO_2 the customers can save by replacing the remaining (non-energy-efficient) windows. The first reports show that a homeowner can save anywhere between 200 and 1000 dollars per year and reduce CO_2 emissions by up to 1.5 tons per year by replacing their windows.

This example shows that Codan through their CSR activities can help to change or influence people's behavior in a more sustainable direction so that they avoid injuries and become more climate conscious. On a short term, Codan is working actively to improve adaptation to climate change and to guard against extreme weather, for example, through climate research, sharing of data at the industry level, and dialogue with local authorities on climate adaptation plans. In the long term, Codan is working with WWF and other organizations on improving the climate and energy optimization. For instance, Codan has developed climate-friendly insurance solutions, and continually works to reduce their own CO_2 emissions. In 2010, Codan measured a decrease of 12% in their CO_2 emission compared to 2007. Likewise, they actively support sustainable energy and are the leader in insuring offshore wind turbines. In addition, Codan has developed a responsible procurement policy where they require that all their suppliers support climate and human rights. Furthermore, their facility management department has to ensure that all their procurements are sustainable and energy efficiency (e.g., through the purchase of energy saving bulbs, recycled paper).

Health is also an important CSR focus area for Codan as the company is a key player on the health insurance market. In February 2011, Codan launched an independent diagnosis insurance, which ensures that their customers can get examined by a specialist within 10 working days, and do not have to wait up to 3 months to get a diagnosis from a public hospital. The safety issue is also central to Codan's CSR strategy. The corporation has access to numerous data about how children accidents happen and has founded the Child Accident Foundation together with Karolinska University Hospital in Stockholm and Odense University Hospital in Denmark through which they implement campaigns for accident prevention. Furthermore, they have presented accident statistics and supplied information about right-turn accidents in collaboration with Danish Transport and Logistics (DTL) to reduce such accidents. One of their latest initiatives is the tool "Driver Profile," which helps to improve safety among truck drivers on the road. A survey of drivers' behavior helps to improve safety and provide more aware drivers.

Among the key learning from integrating CSR into business, Jeannette Fangel Løgstrup mentions, *It is important that CSR strategy supports the overall business strategy, otherwise it makes no sense for the organization and is not sustainable in the long run. CSR can only succeed if it is integrated into the business. How committed is CSR to the business? CSR must live in the organization. We therefore try to break down barriers to colleagues, who may not think that CSR is the most interesting topic. Then we prove that CSR is actually a part of their everyday lives, and that they also play a role, if we are to succeed in our sustainability efforts.*

Consequently, the CEO of RSA has CSR integrated into his bonus targets. That way RSA ensures that CSR is a top priority from the very top to the bottom of the organization.

The CSR director further addresses the specific characteristics to consider when integrating CSR into a service company and mentions that *In a service business, it is crucial to involve the employees in the CSR work since our company is our employees and their benefits. And it is their good ideas to further develop our work with these things. We cannot just replace a machine, and we must change the way you work and the services we offer. CSR must therefore be something they think is important, and that makes sense to them; otherwise, you get them on board. Our product is well in employees, and "product" occurs in the interaction between the employee and the customer. Their willingness and enthusiasm is what motivates and attracts our customers.*

Specifically, Codan has formed an interdisciplinary team, the Nordic Green Team, with the aim to develop climate-friendly solutions for the customers — a team in which WWF is also represented. Furthermore, they organize the annual health week, a climate week, and a volunteer week—so that employees get the opportunity to engage in the CSR themes of Codan and RSA. Another interesting aspect of sustainable business relates to recruitment. The new generation expects companies to be responsible, and CSR is therefore an important element in Codan's employer branding and attraction, competence development, and retention of new generations of employees.

As is visible from the RSA case, the CSR work in the service industry is booming, and within a few years, several major international service companies and especially representatives of the banking industry (e.g., Commonwealth Bank of Australia and Westpac Banking) have moved up in the heavy league of CSR-based organizations. The service industry is dependent of employee knowledge and represents a large percentage of Western countries' GDP. Service companies are not traditionally large consumers of fossil fuels, energy, and water, and therefore do not affect the environment to the same extent as production companies. However, due to their unique industrial characteristics, they affect, engage, and employ a lot of people across the world and interact with customers every day. Banks, insurance companies, and other service companies affect people's lives every day, and the service companies' role as advocates and practitioners of sustainability is therefore central because of their impact on society and their economic power. Many service companies have initially been inspired by literature and practices from the manufacturing industry, since there has not previously been that many other places to look. Yet, during the past decade, an entire wave of studies focusing on CSR in the service industry has emerged. By not being the first industry on the CSR journal, the service industry has in many ways avoided some of the "childhood diseases" and expensive learning curves that production companies had to go through. However, in return they have managed to create an entirely new approach toward corporate sustainability that fits business types without physical production. In summary, the service companies' sustainable business approach includes, to a greater extent than manufacturing companies, an emphasis on philanthropy and cooperation with NGOs where employees' work with sustainable values play a much more important role in the integration of sustainable business. The service industry is a major source of new CSR research and the development of sustainable business as a concept, and will in practice probably be the sector where we will see the biggest CSR development over the next years.

8.3 CSR Integration in Public Organizations

CSR in public organizations is still a relatively new topic, both in theory and in practice. Municipalities and other public enterprises and organizations have only started in recent years to communicate about their sustainability and CSR initiatives. In addition, a smaller percentage of global municipalities have launched a public CSR strategy. CSR is defined in the literature and research from a business point of view, but nevertheless, social responsibility is at the very core of public organizations as they are an integral part of society. Thus, CSR is and will increasingly become part of the public display and communication of sustainable activities that municipalities carry out for society every day (Albareda et al., 2008). Laura Albareda, who is an assistant professor in sustainable strategies at Deusto Business School, have together with her colleagues compared the CSR initiatives and public policies in three European countries—Italy, Norway, and England. The study revealed that the authorities in countries incorporate a common approach and discourse on CSR, and work in partnership with the private and social sectors. For public administrations, sustainability implies an important task of managing multiple sets of relationships, to help develop a win-win situation for companies, citizens, and social organizations. The public sector has several roles to play in strengthening sustainability among its key stakeholders, such as:

1. In encouraging sustainability among businesses and citizens in society
2. Through supporting partnerships between businesses and the social sector
3. By incorporating sustainability in public services, benefits, and productions made/ordered by the public organizations

CSR in public services—both in theory and in practice—is not well described, which initially makes it difficult to identify and present good examples of CSR in public companies. However, one should keep in mind that some of the main objectives of most public organizations is to support social responsibility in society and local communities and to provide help to citizens. Fundamentally, public agencies are responsible for carrying out several sustainable and ethical functions through their work (Steurer, 2010). They assist citizens with health care and nursing, advisory assistance, child care and education, elderly care, and financial support, e.g. welfare benefits, and state pensions, just to mention some. Eventhough these public authorities and organizations basically build on a sustainable values, they should also integrate sustainability across their internal and external functions and activities. In practice, public organizations and municipalities can pursue sustainability through e.g., efforts in reducing waste, electricity, water and fossil fuel consumption, environmental improvements, sustainable employee working conditions (e.g., reduction of injuries, stress and sick leave), as well as green procurement. Particularly, green procurement is a central topic in the state and public organizations' accountability, as these entities initiative massive projects, which require vast amounts of resources – e.g. when building cities, roads, bridges, schools and hospitals.

The lack of communication of public CSR internally and externally is probably due to the fact that government agencies already have a social, ethical, and sustainable role in society. However, public companies also need to prioritize and optimize the level of sustainability (social, ethical, and environmental) in all its actions and start communicating their CSR activities to the public. The unique characteristics of public organizations also influence the proper CSR integration, as they do not have to generate profits, Thus, CSR communication and CSR strategies are not applied in the same way and with the same objectives as among private companies. However, companies and government agencies often rely on each other and have to form alliances and networks to solve social tasks, e.g., in relation to employment initiatives (Beer and Damgaard, 2007) and integration of immigrants and refugees and resocialization of criminals and work integration of handicapped and weakers citizens.

Yet, one could ask why should government agencies and municipalities communicate their CSR efforts when they by definition work with the social responsibility of society? For one, public organizations are role models for their employees, the citizens in the community, and for the society at large. Second, CSR communication on public organizations' CSR initiatives can also be applied to attract new residents and businesses to municipalities and to attract new employees. Many public organizations already work with sustainability. However, it may not be perceived internally as CSR, as the CSR terminology is typically not applied and as CSR initiatives may not structured in a strategic manner or integrated and communicated in a CSR strategy, which makes it difficult to communicate externally.

There is, however, one area of CSR, which in many Western and particular European public organizations and municipalities is well integrated, and that is sustainable and green procurement. The main reason is that the European governments are subject to EU's procurement directives, which restricts the public in its procurement requirements and demands the integration of green procurement policies. The EU's CSR pressure on public organizations has both a positive and negative effect. One the positive side, social responsibility will be integrated in policies as it is requested, but the initiatives and their integration may not necessarily benefit from being pushed forward by directives. Particularly, as the success of these activities in society requires a common interest and understanding of society and the different stakeholder needs for sustainability. In continuation hereof, one can speculate to which extend the public organizations have managed to anchor sustainability down to the individual employee and stakeholder, as sustainability in the public context is ordered from outside, by EU and the state and not motivated to the same extent from the inside of the organization.

The disadvantage of public organizations not communicating their CSR initiatives publicly relates to among other the citizens' lack of knowledge about their municipality's sustainable activities, which can help brand the municipality and attract more investors, businesses, and citizens. Another disadvantage has to do with the lack of learning and knowledge sharing across public organizations and municipalities in enabling and supporting best practices and sustainable results in/for society. Also, the lack of knowledge dissemination to smaller businesses and company owners of how the municipality can assist them in becoming more sustainable as part of the community.

The issue to consider in communicating and making CSR visible , is of course that public display of sustainability will be subject to some level of scrutiny and criticism. However, criticism is not only to be perceived as negative, as it can assist in creating new and more effective sustainable developments and improvements to the existing public practices and services in society. With the lack of communication and dialogue of public organizations' CSR activities, a lot of knowledge and joint, sustainable value creation is lost. That is also why a clear recommendation to municipalities and other public organizations is to communicate their CSR initiatives to the public and to actively engage their stakeholders in defining, prioritizing, and developing new, sustainable initiatives that support the well-being and livelihood of citizens, companies, and other key stakeholders in the community. Fortunately, more and more municipalities pursue and communicate a strategic CSR approach, which the next case with Aalborg Municipality is an example of illustrating how sustainability can be built into public organization's various functions and tasks.

8.3.1 Case Example: Aalborg Municipal

With the adoption of Aalborg Charter in 1994, the municipality's responsibility to pursue sustainable development was defined. Aalborg Charter was initially based on the Rio Conference in 1992 and the Brundtland Report of 1987, and defines the sustainable development needs of the local communities and authorities based on the motto from Rio, "Think globally, act locally." Aalborg Charter was extended in 2004 by the Aalborg Commitments, which places additional obligations on local and regional authorities. Besides Aalborg Municipality, around 650 other municipalities and regions in Europe have signed the Aalborg Commitments and use it as the foundation for their sustainability efforts.

Aalborg is Denmark's third largest municipality and employs around 19,000 employees. The municipality's CSR strategy emphasizes climate and energy and has selected "a healthy municipality" and a safe working environment as two key focus areas of social responsibility . Since 2008, Aalborg Municipality has as workplace developed a common management system for working environment designed under the common OHSAS 18001 standards. It states clearly what good leadership and involvement, motivation, and minimizing absenteeism are—just to name some of the key elements. The system is implemented in the municipality's seven administrations and includes skill development and internal auditing of working environments. A close cooperation between management, union representatives, and safety representatives is necessary in ensuring the link between the solution of the core function (service to citizens) as well as the importance of a healthy working environment. "Well-being and productivity" are one of the municipality's strategic focus areas and has been announced for all the municipality's 900 managers at a management conference in January 2011. The activity has also included targets of e.g. reducing the short-term sick leave by one day per employee. In continuation, of the healthy working environment, the municipality has integrated a MED agreement that puts the focus on the involvement and participation of local authority employees. Furthermore, the municipality has integrated system safety management (SAM). Via SAM, an international certifiable level of working environment is established corresponding to OHSAS 18001 standards. Consequently, an internal auditor corps has been established, consisting of special trained employees, who perform audits of the working environemtn across the municipality's many jobs and functions.

The municipality aims to be the healthiest municipality in Denmark and have therefore launched various initiatives in relation to health, the chronically ill and in relation to children. These initiatives have generated positive results and over 4 years, they have reduced the number of daily smokers by 22%. They have also visited all the schools in the 7th grade to talk about health and to initiate health promotion. Specifically, they measure and evaluate the improved fitness through fitness ratings. Furthermore, they have introduced initiatives in relation to the rehabilitation of patients with chronic diseases.

CSR is described in detail in the organizations' sustainability strategy, which is broadly anchored in the organization of the municipality and across all municipal departments and sectors. A key emphasis in the integration of CSR, is that the municipality actively involves employees and users in the process In the integration of the CSR strategy, the individual administrations and sectors have also integrated sustainability activities in relations to what made sense for each area. Some of these CSR initiatives are among other: green procurement, organic food, sustainability assessment of plans, green accounting, environmental management, and climate-friendly construction.

Today, all the municipal utilities are environmentally certified with ISO 14001, and green accounts are prepared for a large number of companies. In relation to energy management, the municipality has integrated energy meters to make employees and the different functional areas more aware of energy consumption. Specifically, representative from the municipality have visited all the schools in the municipality implemented training programs with service managers, providing them with knowledge about how to change toward more energy-efficient approaches. In addition, 160 kindergartens are about to undergo the same process.

Furthermore, systems have been implemented, for example, automatic switch-off, heat control, and better technical installations in the attempt to lower energy consumption across the municipality and its sectors and functions. Climate and CO_2 emissions are important elements in the municipality's CSR strategy, and they have therefore volunteered as a "climate municipality,". This implies that the organization at a minimum has to reduce their CO_2 emissions by 2% per year. Previous year, the municipality could show a reduction of 5.3%.

The municipality's procurement and requirements for green and ethical procurement are known throughout the organization. Aalborg's procurement policy has the following objectives regarding sustainability:

"Goods and services purchased for Aalborg must be sustainable and must meet minimum wages and our standards for working conditions, health and safety, environment, climate, democracy and social responsibility. Goods and services from companies that are certified under EMAS, ISO 14001 and

OHSAS 18001 or similar certified systems must be prioritized, and environmental and sustainability-labeled products will be preferred where it is technically, qualitatively and economically feasible."

There are specific requirements that the procurement policy has to follow in ensuring that all products purchased by the municipality have been environmental and sustainability assessed. This applies to e.g. organic food, sustainable wood, chemicals in products, eco-friendly vehicles, energy requirements for products, environmentally friendly cleaning products, etc.

In Aalborg Municipality's procurement policy, the following assessments of products and services are carried out:

- Substitution of environmental hazardous substances
- Environmental impact
- Climate load
- Working load
- Ergonomic consequences
- Environmental and climate policy and accounts at prequalification
- Environmental Certification, ISO 14001, OHSAS 18001, or the like

CSR is rooted in different ways in HR and administration, where the municipality's HRM has emphasized the development of the municipality's 900 managers with a focus on creating an attractive workplace with value-based management, where employees are involved in problem solving and key decisions about their work. The municipality has a strategic focus on "social capital" and the importance of practicing trust, cooperation, and showing fairness in their daily work. Competence development and skills training of municipal leaders at diploma and master's level are paramount. As an example hereof, the municipality carried out a conference of 2,300 executives, employees, and union representatives in social capital in March 2012.

An example of a CSR activity set in motion to improve the working environment constitutes: "Project—cleaning in dialogue", which aims at enhancing the psychosocial working environment for the cleaning assistants in schools, institutions, and offices in Aalborg Municipality. Lack of recognition, unclear divisions of labor, and little dialogue about the assignments are some of the main causes of mental stress. The project's key initiatives include training of cleaners and managers, workshops, info sessions, local café meetings, and evaluation sessions before, during, and after.

Another CSR initiative is the 3-year project: "Leaders as health ambassadors and promoters of wellbeing", where 90 managers are provided tools and inspiration to work more effectively with health and well-being through, e.g., training, coaching, and sparring, while they undertake to communicate the new insights to 5–7 colleagues. In this way, the municipality expects that between 450 and 630 leaders are targeted by the project. The leaders further have to undertake the launch of at least 1–2 health promotion initiatives in their own group of employees.

In the administration, the municipality implements several CSR initiatives, for example:

- Print on both sides of the paper (standard configuration)
- All printers set to black and white
- Multiple video conferences (since they minimize transport)
- Introduction of iPads (as they minimize print significantly)
- Green IT, including environmentally friendly operation of the server, automatically turns off PCs

A significant element in Aalborg Municipality's success with CSR integration consists in their diverse collaborations. As part of the the municipality's objectives and to the inspiration of the private sector, Aalborg has focused on cooperation with businesses and industry on the development of environmentally and climate-friendly products through public/private partnerships and networking. For example, the municipality has introduced the following initiatives in cooperation with private and public partners and stakeholders: Green stores, Network for sustainable business development, Flex-energy, North HUB, the Green Board, Network for sustainable agriculture, etc.

Thomas Kastrup-Larsen, who is heading Health and Sustainable Development in Aalborg Municipality, stresses particular one important learning from integrating and working with sustainability across the organization: *the need to measure, so you can follow the development and see the results. You have to set objectives that are visionary, but people must also understand them and have the perception that the targets can be reached, otherwise the whole thing becomes a little "toast speech-like."*

In relation to the unique characteristics of municipalities' work with CSR, Thomas Kastrup-Larsen emphasizes, *In the public sector taking a social responsibility is basically rooted in the solution of many service jobs and in many ways a natural and integral part of the way we work. The public sector may have had a long tradition of focusing on CSR, but without using precisely that concept. In the future, we should focus more on highlighting the many experiences and results – and thus reap the recognition and gain the benefits – example by attracting competent employees and new residents, investments and jobs to the municipality. This means more visibility and requirement to incorporate CSR in the financial statements and in a comprehensive CSR strategy in order to achieve a higher degree of consistency and focus on the implementation of CSR internally and externally.*

CSR in public companies and organizations have long not been received the proper attention and is often undocumented. With several municipalities' clear statements on sustainability through Aalborg Commitments, a new interest practices and a new research field of CSR has emerged. Due to the unique characteristics of public enterprises' special management structure and business model, without profits and where directives come from state or EU, a new CSR model targeted at public organization is requested. Little help and inspiration can of course be gathered from the private service industry., However, ultimately public institutions must develop and incorporate the CSR approach that fits their unique characteristics and culture. Specific sustainability requirements are of course presented by the state, governments, and the EU in the form of green procurement. However, there are still a lot of "gray areas" in public organizations CSR that has not legislated yet. In these cases, municipalities must choose side and target their CSR on the organization's unique opportunities and challenges. The potentials for CSR in municipalities are massive. For one, public CSR can contribute to the positive reputation and corporate branding of municipalities and other public enterprises, and in new ways that may help in attracting new and skilled employees, citizens, businesses, investors, and other stakeholders than before. It will therefore be interesting to follow the trends in public CSR over the next coming years.

8.4 CSR in Small- and Medium-Sized Entities (SMEs)

The CSR literature has so far paid disproportionate attention to larger American organizations (Spencer and Rutherfoord, 2003). Consequently, the theoretical foundation of CSR is fundamentally American and strongly reflects the Anglo-American business context (Crane et al., 2008). Although there is a growing interest in CSR among SMEs, many analyses confirm that large organizations are still the main developers of CSR strategies (Hernáez et al., 2012).

Nevertheless, SMEs are particularly sensitive to CSR issues. For one, the relationship with local authorities and communities are closer and more direct than of larger organizations. Second, due to size smaller economies and potential lack of branding, attracting the right resources and investments, can be a challenge. Crane et al. (2008) emphasize in their study three assumptions about CSR in SMEs:

1. CSR is rather informal and ad hoc. Trivial issues are usually ignored. Instead, the topics of interest are issues considered critical in a given context. That is, issues that are potentially embarrassing or scandalous.
2. SMEs go under the radar of wider society.
3. CSR in SMEs is targeted at building good personal relations, networks, and trust (Spence and Schmidpeter, 2002).

It can be argued that this third assumption encompasses an internal reasoning of CSR, which is in contrast to large corporations that reason externally in their engagement in CSR (Morsing et al., 2008).

CSR in SMEs is also claimed to be more value-driven, as many SMEs are typically owner-managed firms where ownership, control, and decision-making is performed by the same person (Jenkins, 2006). The personal and ethical values of the SME owners and the employees driving the motivation to implement CSR, are key elements to consider for the SME in integrating CSR activities that are built upon these values or responds to them (Verhaugen, 2003). Jenkins (2006) suggests that an internal drive rather than external pressure is the main driver for CSR in SMEs. Finally, personal relationships are often both preferred and predominant in SMEs, and research suggests that SMEs' absence of bureaucratic controls could enhance the openness and trust in business relationships (Spencer and Rutherfoord, 2001). Apart from these general assumptions, it is important to notice the differences among SMEs based on country of origin. CSR is an idea that needs contextualizing in the relevant social context.

Micro-, small-, and medium-sized enterprises (SMEs) constitute 99% of companies in the EU. They provide 2/3 of private sector jobs and contribute to more than half of the total added value created by businesses in the EU. Nine out of ten SMEs are actually micro enterprises with fewer than 10 employees. Various action programs have been adopted by EU to support SMEs, such as the Small Business Act, which encompasses all of these programs and aims to create a comprehensive policy framework. The Horizon 2020 and COSME programs have also been adopted with the aim of increasing the competitiveness of SMEs through research and innovation, while providing better access to financing for SMEs (Schmiemann, 2008, www.europa.eu).

These are some of the reasons why SMEs have become a popular research area, where more and more researchers and practitioners discuss the importance of CSR in the development of SMEs (Morsing and Perrini, 2009). However, there is a huge difference between the way in which SMEs and large enterprises integrate CSR. Research shows that formal CSR strategies are most common among large companies, whereas the integration of CSR into corporate strategy and behavior has a more informal mark among SMEs (Perrini et al., 2007; Russo and Tencati, 2009). For one, SMEs rarely use the CSR terminology or characterize their CSR activities as sustainable (Russo and Tencati, 2009), which can rightly be characterized by the concept of "tacit CSR" (Perrini et al., 2007). This also makes it harder to compare CSR across company size, as SMEs do not document, disseminate, or communicate their CSR-related activities in the more formal and strategic manner of most large organizations.

According to Professor Mette Morsing (2006: 3), CSR is much more institutionalized in large companies than in SMEs, and despite CSR's popularity among many different types of companies, the specific content and results are still questionable and indicates that CSR is only semi-institutionalized in SMEs. Several of the public and EU-based CSR initiatives have helped in increasing the focus on the need for CSR in SMEs. However, a Gallup survey of Danish SMEs from 2005 showed that 75% of small- and medium-sized companies were involved in CSR activities, but only 36% of them had communicated their CSR activities externally (Gallup, 2005). It can be argued that many reporting and certification systems are difficult to administrate and organize for small businesses. The institutional pressure often leads to environmental activities of mere compliance, which in turn have been found to have an insignificant impact on a firm's business activities (Darnall et al., 2008).

CSR is often conceived and applied as a concept in relationships to large companies, which are of such size that they affect the society and the environment significantly, in relation to economy, environmental effects, and employment. Individual SMEs rarely affect the environment to the same extent. However, the mass of SMEs does. Another issue in CSR integration among SMEs relates to the fact that they do not have the same resources to carry out CSR activities as the large companies, which often operate with million dollar CSR budgets. Therefore, many SMEs apparently ignore CSR, as owners and managers of small- and medium-sized enterprises mistakenly have the conviction that CSR requires large investment and only makes (business) sense in the long term. Research reveals that the types of CSR practices that produce a cost reduction effect, as well as being a positive change in social and environmental terms, are those that tend to be most consistently applied by SME owner managers (Ilomaki and Melanen, 2001). This points to an obvious conclusion that it is the cost reduction element of the sustainable behavior change that is motivating, and that the environmental benefit is merely a positive byproduct (Williamson et al., 2006).

CSR as a concept is something that most SMEs do not really relate to although many SMEs actually have lots of sustainability built into their business practices. However, their lack of more targeted CSR actions and a CSR strategy dilute the potential, positive business effects and performance impact as their sustainability initiatives are less visible to their stakeholders and the society. More and more companies, especially the larger organizations, require that their suppliers and partners have a sustainable approach toward production and business (Morsing and Perrini, 2009). It therefore matters to what extent SMEs carry out and communicate CSR-related activities as it can potentially determine whether their company is selected over another as a supplier, subcontractor, partner, and/or workplace.

Theory and practice reveal that CSR integration in SMEs is dependent on the owner/director of the company and their ability to integrate sustainability in the existing business, while making it relevant for employees. Likewise, the fundamental values play a significant role in whether CSR becomes a common mindset or just a casual activity among the employees. CSR therefore has to be linked to the company's values to make sense to the organization. However, the challenge is that not all SMEs have worked explicitly to map their values. Studies in French SMEs found that the major driver of integrating CSR into business strategy related to the beliefs and values of top managers, with economic motives coming in only at second place (Hudson and Roloff, 2010).

Thus, integrating CSR without foundation in the company's values is not recommended. This does not mean that SMEs must undergo large and complex value processes. However, it is beneficial for management in cooperation with the employees to explicitly identify the fundamental principles and sustainable values that the company wants to work from—both internally and externally.

The weighing of the informal and formal approach toward CSR integration in SMEs presents another important strategic choice to consider. If the intentions and strategy of sustainability are primarily informal by nature, then it can also be difficult to explain and communicate explicitly outside the company. At the same time, a formal approach, where the company's

CSR strategy is disseminated externally, obligates the company and makes its sustainable actions visible. "Green hushing" is a popular term for the lack of communication of CSR to stay under the public radar, and this approach is quite common among SMEs.

However, the recommendations are still that SMEs should make CSR visible, explicit, and focused for it to make sense for employees and other stakeholders and to be able to reap the business benefits of their activity (Perrini, 2006).

More recent research reveals that a growing number of SMEs are becoming more confident in being environmentally responsible and communicating about it. Particularly, as they experience that the publicity can help win customers and assist in attracting and retaining staff. It appears that cultivating close relationships with workers and the social, business environment ensures collective action through increased confidence (Murillo and Lozano, 2006), more satisfaction among employees and an improvement in the company's image (Longo et al., 2005).

Furthermore, the integration of sustainable business can open up new markets, new business and investment opportunities, as well as new collaborations. Thus, CSR must be incorporated in the existing business and in the services that the company provides and in the products it manufactures, as well as in the new innovations and sustainable products that customers demand. Basically, the sustainable initiatives of the SME have to make sense in the context in which the company operates. In reality, it is really only the creativity and capabilities of the company combined with the customer requirements and demands that set the boundaries for what the SME can achieve through integration of CSR. The next case is based on the SME, Rynkeby, a medium-sized company manufacturing fruit juices, and reveals how CSR can be embedded and integrated in business and throughout an SME organization through the active involvement of employees.

8.4.1 Case Example: Rynkeby

Rynkeby was originally founded in 1934 and started as a cider producer in a small town in Denmark by the name, Rynkeby. The contemporary history of Rynkeby foods' is a tale of mergers and acquisitions over the past 20 years. Today, Rynkeby is Scandinavia's largest producer of fruit juices and was owned by one of the world's largest dairies, Arla Foods, for a number of years and was recently acquired by the German juice manufacturer, Eckes Granini. In a year, Rynkeby Foods produces more than 150 million liters of juice, which is sold primarily in Denmark and Sweden. The company has 193 employees and created in 2015 a turnover of $180 million.

The strategic CSR activity started at Rynkeby in 2009 with one staff member (a quality and communications manager), who was provided with the core task to launch a project on CSR while mapping the company's sustainable activities. First, the existing CSR-related activities were identified. Second, further knowledge and inspiration were sought from various CSR websites, national guidelines, and Global Compact. In practice, it became apparent that Rynkeby had always worked with CSR to some degree, only without putting the CSR label on. Particularly, optimization of resource consumption and waste had always been a focus area right from the very start of the company.

The former CSR responsible, Carina Jensen, says that *I chose to regard CSR as a potential win-win situation for all parties. I first looked at what activities made sense and created results for us, our customers and the environment – but also with a focus on economic performance, as we are not a philanthropic company. However, in practice, CSR and a positive bottom line go hand in hand. Environmental improvements often provide savings in resource consumption and increased employee satisfaction, and improvements in employees' health means fewer sick days.*

Rynkeby based their CSR strategy and activities on the triple bottom-line and the three p's: profit, people, and planet, while adding health as an extra focus area. Health is the focal point of Rynkeby's sustainable business as it makes perfect sense in a company that produces healthy drinks based on fruit and vegetables. Due to their CSR activities on health, the company won the price for healthiest company in their region in 2010.

Health is an important part of Rynkeby's business-driven CSR as it also makes good business sense to have healthy employees. This strategic CSR focus has generated a number benefits for the company leading to fewer days of employee absence due to illness, far greater and measurable job satisfaction and ultimately a more efficient workplace. This is again a practical example of the economic and health/employee rationale that fitting perfectly together. The financial business effects of Rynkeby's CSR initiatives also make it easier to convince management, employees, and other stakeholders about the importance of such CSR activities. As the company is a SME, they do not have the resources to go in depth with all CSR activities, so they regularly have to assess and determine which sustainable activities are the most important to prioritize.

An important element of Rynkeby's successful CSR integration relates to the fact that the responsibility for the four main areas (people, planet, profit, and health) is distributed among top managers and directors of Rynkeby's leadership team.

Another success factor of Rynkeby's integration of sustainable business relates to employee involvement and the fact that their CSR has emphasized the wellbeing and healthy of the company's most priced resource, their employees. As the CSR director states *as an SME located in a small community, attracting and retaining employees become a core HR activity.* As an example hereof, Rynkeby created health groups across departments that are actively working to create a healthy workplace and healthy employees. The company has also carried out campaigns for exercise; it provides access to sports coaches and inspiration seminars on health and wellness, offering slimming course and stop-smoking courses. Furthermore, they have established their own 'fun run' and nearly 100 employees (of its 193 employees) participated in the national DHL relay. In addition, the employees are also offered a health check and can get a calculation of their "BodyAge." All of these activities are measured, and its Rynkeby's goal to reduce the overall BodyAge of the company's employees by 3.8 years. In so doing, they have therefore included various competitions for employees to get in better shape, lose weight, and stop smoking. The company has already seen great effects in relation to increased attendance, fewer sick leaves, and increased job satisfaction.

Another major CSR focus area for Rynkeby is the environment. The company is located in a small community, which the company is highly dependent of. Thus, social responsibility in Rynkeby has to emphasize the local aspect. In this area, Rynkeby includes efforts to reduce waste, and they transfer their excess heat back to the heating plant of the municipality that Rynkeby is part of. Employees receive reminders about how they can improve the environment. For example, operators must remember to sort packaging, limit travel, turn off the lights in the departments when they leave, and so on. It must make sense to be carried out by all the employees of the company. Furthermore, Rynkeby also collaborates with local suppliers and subcontractors in the area to improve the environment and the local community. Specifically, the company wants to reduce energy consumption and waste water by 10%, raw material consumption by 30%, packaging waste by 50%, and CO_2 emissions by 10% per produced liter.

Rynkeby has also integrated CSR into their communication and work with "fair speak," which basically means that they provide the consumer with accurate, complete, and truthful information. Similarly, the company communicates the company's CSR initiatives through its staff magazine, which is published quarterly. Rynkeby also regularly organizes canteen meetings where employees are informed about CSR efforts and the results are made available to employees on the company's intranet. In the spring of 2011 Rynkeby published its first CSR report, which was sent home to all employees and this practices has been continued.

Rynkeby has specifically focused on internal communications as management emphasizes the necessity of CSR "living" among the companys key stakeholders, their employees. Rynkeby's corporate communication is known for one thing in particular, their Team Rynkeby cycling team. The employee-driven idea of a company cycling team originate more than 10 years ago, when at a Christmas party 8–10 employees decided to go to Paris during Tour de France to donate money to charity. This tradition has grown into one of the largest cycling events in Scandinavia. This year, the company reached over 500 participants, with almost 40 employees from Rynkeby and $5,5 millions being donated to the Child Cancer Foundation. The event generates so much positive publicity both nationally and internationally, which has strengthened Rynkeby's corporate brand tremendously. Likewise, employees get great exercise at the event as well as during the training beforehand, which helps to create unity across different departments in the company.

What is the challenge of working with CSR in SME? Carina Jensen from Rynkeby states, *The biggest challenge has been that I was given a task by management, which management probably did not initially known the extent and significance of. The presentation was, "we must also have CSR," but the problem has been that the management team did not have great insight into the various strategic CSR directions to choose from. Top management's different views of what CSR should be and not be, has another challenge. In the sales department, child labor and code of conducts are what customers are talking about. Others thought that the CSR focus should be on the environment. So we have had many discussions about the contents—before the work began.*

In relation to what recommendations Rynkeby can give other SMEs that are considering CSR, Carina suggests the following:

- Start by defining what is important for your business. What is your business built on and what CSR activities come most natural in relation to the core business, vision, mission, and values?

- Map what the company expects to get out of CSR in concrete terms? How is it measured? What is the need really?
- Select only a few focus areas—such as the three p's—and select the most important areas. Set goals which all can work to achieve.
- Management should get a course in CSR before the development of a CSR strategy to ensure a common understanding of the concept.
- Keep track of the good examples during the process and not just at the end of the year.
- Use enthusiasts in the company to drive the process and the good stories forward.
- Create alliances and networks for sparring and inspiration.
- Resources have to be deployed in order to prioritize and perform on CSR.

In summary, Rynkeby did not change that much in their existing business and practices, but they systematized their activities. Their CSR activities have now been put into the appropriate strategic CSR focus areas under the CSR umbrella—making them visible to the employees in relation to what to do. As the case from Rynkeby illustrates, SMEs have often already worked sustainability, they just call it something else and typically regard sustainable business as common sense.

CSR in SMEs has turned out to be a research area in its own right due to the potentials of integrating CSR into SMEs. The fact that a large percentage of the world's GDP and jobs are generated by SMEs emphasize the key role of SMEs in national and global sustainability and there is need for more attention. Meanwhile, there is a huge untapped potential in using CSR as leverage for these organizations, their growth and development, services and collaborations. Practice reveals that SMEs do not apply their sustainable business sufficiently in their communication. This is a pity as there are great business potentials to be achieved in relation to attracting talented staff, co-operations and investments that help to create competitive advantages. The main task of SMEs is to get their existing CSR-oriented activities structured and focused to such an extent that it can be communicated to the company's stakeholders and used to create new growth opportunities and competitive advantages.

As this chapter shows, there is a big difference between the way in which CSR is incorporated and used among different industries and company types illustrated by the cases presented in this chapter. However, there are also several similarities in the different company's various approaches to sustainability when you compare across the cases. All the interviewees emphasize how important it is that CSR makes good sense for the company's business, products, and stakeholders. The conclusion of this chapter must therefore be that you cannot copy other companies' CSR practices, instead you have to customize the content and form to the unique characteristics and culture that characterizes your specific organization.

9

CSR in Society and in the Future

The society and the stakeholders of a company play a key role in CSR, as companies can no longer act as isolated business silos. Instead, they have to act as interactive units as they influence and are being influenced by the surroundings for better or worse. However, what is society really and how can a company actively apply society and its stakeholders in relation to improving their sustainable business and corporate sustainability? The definition may be broad, and in this book, the society influencing a company really contains all possible stakeholders. This is also why the definition of society by Freeman (1984: 189) is applied in this context:

Any group or individuals that can affect or be affected by an organization's goal setting.

In practice, this means that society contains all the stakeholders that may have an interest in and/or interact with the company. And that is basically all.

Even though you may not consider yourself, another individual or unit, as a stakeholder to a company at one point, you, or that individual or unit, may become a stakeholder the second that company e.g. decides to move into your community, when you, a friend or family member gets hired by that company, if you use the company's products and services, and/or if the company's news or scandals affects or provokes you.

Most people are therefore stakeholders to several companies and organizations without knowing it or being actively involved. This also implies that we as individuals through our stakeholder roles can influence many companies and even can act as a kind of protector and/or ambassador for society and the environment. In addition, we can have multiple stakeholder roles to perform simultaneously. If, for example, we live in the next door and are employed in the company, we constitute a stakeholder of the community as an employee. Through these roles, we can exercise our power and air our viewpoints, which may affect the company and the way it does business, the way it produces and sells its products and services, and/or the way it manages its employees.

By making use of our "voice" in society, we can apply different venues and channels of communication such as trade unions, environmental organizations, municipalities, political organizations, social media, and various complaints boards to help create a better society.

The concept "social" in CSR has always been difficult to define and has lacked direction as to who the company is actually responsible for. With the concept of stakeholders, corporate social responsibility is being personalized, because it is limited to the specific groups or individuals to whom the company is responsible and should therefore consider communicating its CSR to (Carroll, 1991). Stakeholder management or management for/of stakeholders has therefore become a key issue in effective facilitation and management of sustainable businesses. The major tasks in stakeholder management are to describe, understand, and analyze the stakeholders of the company and to identify how they should manage stakeholders. In so doing, Carroll (1991: 44) suggests five questions, which the company should ask in clarification of its stakeholders:

1. Who are our stakeholders?
2. What are their efforts?
3. What opportunities and challenges are presented by our stakeholders?
4. What social responsibility (financial, legal, ethical, and philanthropic) does the company have toward to these stakeholders?
5. What strategies, actions, or decisions we must take to best manage these tasks?

Earlier in the book, the model by Morimoto et al. (2005) was suggested in mapping the stakeholder activities. However, in mapping the stakeholder, another matrix is suggested, which is the (original) stakeholder matrix by Archie B. Carroll:

Table 9.1 Mapping the responsibility of the company

Stakeholders	Economic	Legal	Ethical	Philanthropic
Owners				
Customers				
Employees				
Immediate environment				
Competitors				
Suppliers				
NGOs and activists				
Society				

Source: Carroll (1991: 44).

The model assists in providing an overview of all the company's key stakeholders and in mapping the specific economic, legal, ethical and philanthropic responsibilities that the company has toward the specific stakeholder groups. The model is a template for the company to fill out. However, to get the full potential of the model, company representatives should use the model as a tool for stakeholder dialogue in mapping the stakeholders' specific expectations toward the company. Because if what the company views as their responsibility and what the stakeholder sees as the company's responsibilities are not aligned and alike, then there is a potential for future conflicts.

9.1 Society as Watchmen

Society's increasing attention of businesses' social responsibility implies that companies can no longer hide or get away with greenwashing. In particular, social media and the global press play an essential role in public debates and in society's and stakeholders' assessment of an organization's true corporate sustainability. The global opportunities for open debate through Internet and social media provide the platform for easy, global, and quick dissemination of scandals and "revelations" of "not so green" CSR stories and corporate communications. This puts pressure on more coherent and sincere business communication. For example, when in the late 1990s, it was revealed that IKEA's children's furniture was manufactured using child labor from India, a public scandal exploded. The reason why the story took of in the media probably had to do with the public image of family values that IKEA had sent out to its costumers, and which did not fully harmonize with their use of child labor.

Another public scandal arose when it was revealed that a selection of US prisons applied the tranquillisers manufactured by the pharmaceutical company, H. Lundbeck, in the injections given to prisoners sentenced to death. H. Lundbeck obviously did not develop this medication with the intention to kill or with this use in mind and had not been contacted or asked by these prisons about the use of their medicines in lethal injections. Yet, the story took on in the media anyway. Although H. Lundbeck was not to be blamed, the incident still provided poor press for the company. In this case, the company has to their rights while still taking responsibility for the company, the value chain, as well as the damage that the products can potentially cause—also when used incorrect or unintended. Such cases give rise to some serious investigations and a public assessment of the companies' sustainable business and responsibilities. In practice, the company's responsibility are extended to the very last person that handles the product and that also goes for the many subcontractors that the suppliers (and not the company) have engaged in collaboration with. So despite good intentions and code of conduct agreements with the (direct) suppliers, it can be quite difficult for a company to ensure that their products are manufactured, sold, and used correctly and that all the subcontractors and other stakeholders involved are also behaving sustainably, ethically, and environmentally correct.

In the cases of IKEA and H. Lundbeck, one could ask the question: who should really be held liable? Is it the Legal department or the Communications department who should be blamed? Or procurement as the supplier's subcontractor did not respect the agreements made with the supplier? The fact is that no matter what the company does and how much they try to prevent these scandals, the ultimate responsibility of corporate sustainability always runs back to the company at least in the public eye. The company's good reputation will be tarnished anyway, no matter how much the company defends itself as the "story is out" and very hard to take back. Therefore, a practical advice is for companies to always carry out a certain level of self-discipline, self-criticism, and self-control across the enterprise and its collaborations, which ensures that the image that the company sends out to society also holds true in reality.

Another practical advice is to establish tools and guidelines for corporate crisis management as most companies eventually will experience some kind of public criticism and/or public "scandal." Effective crisis management requires that the company assumes its responsibilities in public, and quickly and efficiently communicates and acts on the case. Press kits can be prepared in advance in case a scandal hits the company. That way the company can establish clear, internal and external guidelines of how these negative press cases should be handled professionally and effectively. As the US stock broker, philanthropist, and self-made billionaire, Warren Edward Buffet, stated: *It takes 20 years to build a reputation and five minutes to ruin it. If you think about that, you'll do things differently.*

Bad publicity and negative reputation rubs off on the company's brand and can in the worst cases result in lower sales and earnings. Thus, the company's sustainability depends on society, its stakeholders and their perception of the company. Consequently, the social responsibility of a company covers all the stakeholders, who affect and are affected by the company and its actions.

9.2 Global Growth Through Sustainable Collaborations

A company's collaborations and partnerships are part of its public image, and they can quickly put both positive and negative imprints on the organization's brand and reputation. Thus, in practice it is necessary to manage, control, and develop the sustainability of partners with which the company cooperates as well. It may in fact imply that the company has to establish joint agreements and alignments of the partners' expectations with clarification of their attitudes and values toward sustainability, In practice, this is often done through the application of codes of conducts as part of the partnership agreement. However, code of conducts cannot stand alone and control is not the solution to everything, for collaborations are about trust and knowledge exchange. Thus, the parties should through dialogue create an understanding of why the code of conduct agreements must be integrated and monitored, and how the company can assist with knowledge and possibly training to support and enforce these agreements and that way develop and facilitate positive collaborations between the partners. In particular, in cooperation with subcontractors in third world countries, there is a big difference between what different nationalities take for granted in relation to social responsibility and how they define this concept in practice. An example hereof is child labor. The laws and values of Western countries dictate that child labor is illegal and cannot be accepted. However, in many third world countries, child labor in families with 5–10 kids is a prerequisite for the families' survival.

Global, sustainable cooperations are therefore also about developing and strengthening the suppliers and their subcontractors through dialogue, knowledge dissemination, and education. Thus, the corporate sustainability of the company is highly dependent on the sustainable growth of its partners and suppliers across the entire value chain. Initially, this approach could appear as "training in western values." However, it should instead consist of developing mutual understanding and development through exchange of knowledge that benefits both partners. The training of suppliers should therefore be adapted to the characteristics of the suppliers and their nationality and culture. In reality, there is a huge potential in the growth generated through sustainable development of global collaborations. However, companies are often left alone with the task, as third world governments neither have the power nor the skills or resources to remove the root to the sustainable issues, which affect the livelihood and the businesses in the specific countries.

Sustainable, global collaborations are not about philanthropy as few companies can survive doing this. It is about creating a healthy and sustainable business with a value chain that benefits the business, the customers, suppliers, and other stakeholders. The company therefore has to decide, which "social battles" it wants to get involved in and where it can make the biggest, sustainable impact on their third world supplier and in the specific region/country, where it does business.

NGOs and environmental groups such as Greenpeace and WWF have long acted as the watchmen of the global environment and conservation, which has shed light on non-sustainable business practices in various organizations and industries over time, including Shell, BP, agriculture, and the fishing industry. These organizations and more industries have 'battled' back and forth legally and in the press with NGOs on many occasions. However, companies should not consider NGOs as the enemy. Instead, they should use their knowledge to overlook at their own organizations and its value chain in examination of how sustainable practices can be optimized and how these NGOs can assist in supporting sustainable business.

9.3 NGOs as Business and Innovation Partners

Businesses and NGOs have a long history of battling against each other across different environmental, social, and ethical issues. And in this context, the UN definition of an NGO is applied, which states that an NGO is a non-governmental, nonprofit, and voluntary organization (UN, 1998). However, during the past decade more and more businesses and NGOs have collaborated on the basis of philanthropy and volunteerism (Austin, 2000). In the wake of global environmentalism, a shift has taken place from the 1980s where especially environmental NGOs assumed a far more visible role on the international political scene (Doh and Guay, 2004). Throughout the 1980s and 1990s, these activist groups employed a new antagonistic strategy toward sustainable business practices in terms of new environmental and social regulative standards (Argenti, 2004; Doh and Guay, 2004).

Recent studies show that there has been a new development in the way NGOs and companies perceive each other and how they get involved in collaborations (Yaziji, 2004; Pedersen et al., 2011). This shift in collaboration patterns and interaction processes provides new opportunities for different types of interaction outcomes. One unique and multi-faceted outcome of business–NGO partnerships is the co-development of sustainable innovation.

The concept underlying most studies in sustainable innovation through business–NGO partnerships is Austin's (2000a) conceptualization of the collaboration continuum (CC), which this study is based on. The collaboration continuum contains different degrees of interaction in partnerships including three stages:

1. The philanthropic stage
2. The transactional stage
3. The integrative stage

The philanthropic stage is characterized by conventional one-way donations to the NGO partner in terms of financial, material, or volunteer resources. The transactional stage emphasizes creation of mutual benefits through reciprocal resource exchange of more valuable resources in terms of sponsorship, cause-related marketing, information campaigns, NGO-initiated codes of conduct, and so on (Pedersen et al., 2011). When collaboration evolves and becomes more integrative, mission and goals overlap somewhat and value creation develops simultaneously for both partners (Austin, 2000a).

More and more companies collaborate with NGOs as part of their CSR development (Kovacs, 2006). These NGOs are private, voluntary nonprofit organizations, which represent social movements and various ethical ideals and standards. Examples here of are WWF, Greenpeace, Amnesty International, Care, Oxfam, and so on. These organizations all vary widely in size, mission, and strategy and are characterized by the specific cases that they are passionate about, communicate, and support. NGOs are also characterized by their network, which can assist businesses in generating new knowledge and business networks and opportunities. NGOs are not based on the same business conditions as traditional profit organizations as they do not have to make a profit due to the fact that they are typically funded by members, sponsorships, and donations. This also implies that they can focus on social and ethical areas, which businesses leave out or do not pay attention to. Generally speaking, their work consists of, for example, information collection, analysis, dissemination of reports, and knowledge and propaganda within the social and ethical matters that concern the NGO. They often have an important role as public opinion makers and in political lobbyism and therefore influence the development of laws and standards that support their cause.

The question still remains, why NGOs and businesses should collaborate and how they can both benefit from partnerships? The businesses can among other apply the NGOs as sparring partners in relation to topics where the NGO is a specialist. Both theory and practice have illustrated that access to knowledge is a major cause of corporate interest in NGO collaborations

(Doh and Guay, 2006). NGOs are important knowledge and network intermediaries, because of their quest for knowledge, their unique networks, and active work in improving specific areas of society. Thus, when a company is about to develop a new product or services, or have to enter new markets or introduce new processes, which may have an environmental or social impact, then NGOs can assist as knowledge partners in answering how these business practices can be conducted in the most sustainable, ethical, and environmental-friendly way. At the same time, NGO can provide, develop, and disseminate knowledge that can help the company in creating new business opportunities and/or products that benefit both the cause of the NGO, the business and its stakeholders. Practical examples hereof have been provided in the cases by RSA/Codan, Hummel, Henkel, and Novo Nordisk as presented in earlier chapters of the book.

At the same time, the political correctness of NGOs rubs off on the companies that cooperate with them. For example, if a company chooses to develop a product for children with assistance and knowledge from, for example, Save the Children, then the products developed through this partnership will most probably appeal to families and strengthen the company's reputation as an organization, that cares for children. However, the application of NGOs in product development and PR campaigns has also received some bad publicity in the public media, as many stakeholders consider these collaborations as NGOs sleeping with the enemy (Burchell & Cook, 2013).

However, the political power that NGOs possess can also help to support causes that companies and the NGO are passionate about. For example, many environmental organizations are very positive about non-fossil fuels, and manufacturers of wind turbines, solar energy, and other non-fossil energy sources can therefore apply the statistics and reports produced by the NGOs as knowledge assets and in their marketing to promote their products on the market as well as in the political and public debate. There is therefore a unique potential of partnerships between NGOs and companies, primarily in relation to knowledge, networks, and product development, but also in relation to PR and communication (Doh and Guay, 2006).

9.3.1 Charity and Business Can Go Hand in Hand

Both theory and practice have emphasized the business potentials of companies applying charity and donations to NGOs in their corporate communications. The research within this field has uncovered the business case in terms of sales increase, consumer awareness, product differentiation, increased reputation, public relations (PR), and employee loyalty (Lodsgård & Aagaard, 2016). Recently, cause-related marketing has become extremely popular as one of the most emerging segments in companies' marketing and the main driver toward philanthropic partnerships (Austin 2000; Porter and Kramer, 2002; Neergaard, 2009; Pedersen et al., 2011). It appears that a shift has taken place, where former conventional "checkbook" philanthropy has now moved from companies' charity budgets into marketing budgets linking the company's products with sustainable causes (Austin, 2003; Vogel, 2005). Other business benefits of philantrophy stress that when employees are tightly embedded in community and charity activities, it is more likely for them to stay loyal to the company (Austin, 2000).

More and more organizations are connecting charity and business. The collaborations and donations are typically targeted at the established and well-known NGOs, which are among top 50 in the world such as: WWF, Red Cross, Partners in health, Danish Refugee Council, CARE International, Médecins Sans Frontières, and at the same time supplemented with new and/or more niche-oriented NGOs and charities such as AARP and Helen Keller International. Common to these NGOs and charities is that they are based on some form of philanthropy and the desire to improve the conditions and circumstances of a particular focus area and/or group of stakeholders.

Charity and business can go hand in hand. However, in creating a meaningful donation, companies have to identify the charities and NGOs that can potentially support the development of their business, help the company in pursuing their CSR strategy, and that also make good (business) sense to their stakeholders. Thus, a strategic fit between the charity and the company and its business has to be identified or else the charity can potentially have a negative effect and cause confusion or suspicion of 'greenwashing'.

If, for example, a company has selected employee health as a strategic CSR focus area, then charities and engagement with an NGO like WHO would make great sense. To create a logical connection between the charity that the company supports, the business that the company is in, and the core products that it offers is crucial to a coherent CSR strategy and CSR communications involving business-driven charity activities. With the specific choice of NGO collaborations and donations, companies can stress key strategic priorities, which again can help to support the company's corporate brand and CSR communication by showing the company's corporate values in action.

9.4 Where is CSR Heading—CSR Version 2.0 or 3.0?

Throughout the book and in the company, cases presented, the practice and theory have pieced together an overview of how CSR is and can be integrated and how sustainable business is facilitated. All of which outlines what is common sustainable business practice today and indicates where we may be heading next. The company cases clearly indicate that CSR is already well-established in many, particularly large companies, whereas many SMEs are not quite as formalized or communicating as effectively about their CSR initiatives (Morsing, 2006). CSR will in the next year with all likelihood be rooted more in corporate business strategy and functions, as it has to be part of business to survive over time and fiscal crises. In practice, this means that CSR is likely to be embedded in corporate practice and business to an extent that rules out separate strategies and departments emphasizing CSR. In time, CSR will no longer be something companies talk about, but something they live out every day as a natural part of working and doing business across the company's global or national value chain. Private companies can obviously only survive if they generate profit. They cannot live on philanthropy. However, they can run a good, responsible, and sustainable business that takes care of the society that it is part of while trying to eliminate and/or minimize the negative environmental and social impact of its products and productions. The sustainable approach to business is a necessity if the company wants to create and withhold a strong brand, and this again can assist in attracting talented employees, new customers, investors, and partners.

The ability of businesses in generating profits will still be important. However costumers, consumers, the authorities, and society will make greater demands on how these profits are generated and how they are applied to foster development of the company's value chain, partners, and surroundings. It is possible that consumers will begin to raise counterclaims for the companies that have benefited the most and have had the highest earnings, for example, by requiring that these companies "giving back more to society," either through donations, third world development program, or opportunities for marginalized employees.

The increase in sustainable businesses will also increase demands for training and educations in sustainable management as well as for skilled employees and managers, who can facilitate sustainable business across the organization.

An interesting business development that is already visible in practice now is the growth and combinations of sustainable and unsustainable businesses integrated under one corporate umbrella. An example hereof is the oil industry, which also invests in solar energy plants and wind energy turbines. Another interesting development is related to the business model innovation that companies undergo to be able to pursue the Bottom of the Pyramid (BOP) markets, which consist of poor, third world countries with huge business potentials due to their size. Another business development potential is also visible in companies that pursue development of products, services, and processes through NGOs.

More and more new companies are today founded on social responsibility and actively try to solve or diminish some of the many sustainable, social, ethical, and environmental challenges of the world (such as water shortage; water, air, and soil pollution, waste and poverty), through their business offerings. A few examples hereof are: LifeStraw, which is straw that purifies polluted drinking water, and Groasis Waterboxx, which is an agricultural box that allows plants to grow in desert areas without irrigation or energy use.

The concept of circular economy is gaining increasing theoretical and practical emphasis. Particularly, due to the focus put by EU, which has adopted an ambitious Circular Economy Package. This package includes revised legislative proposals on waste to stimulate Europe's transition toward a circular economy that will boost global competitiveness, foster sustainable economic growth and generate new jobs. Furthermore, the package also consists of an EU Action Plan for the Circular Economy that establishes a concrete and ambitious programme of action, with measures covering the whole cycle from production and consumption to waste management and the market for secondary raw materials. The proposed actions will contribute to "closing the loop" of product lifecycles through greater recycling and re-use, and bring benefits for both the environment and the economy.

Waste management is an interesting trend, as waste is no longer just something that a company has to find a sustainable way to reduce or get rid off. Waste also represents new business opportunities and combined with the circular economy mindset industrial symbiosis can be stimulated—turning one industry's by-product into another industry's raw material.

Some examples hereof are Terracycle, which is a company that converts plastic waste into plastic materials such as dust pins and park benches. Bio-bean, which is a company that converts coffee grains into biogas; and Biotrans, which is a system that collects food waste from restaurants and canteens and converts it into biogas.

Business–NGO partnerships also represent an interesting trend as both can gain from each other. The NGO can live out its political power and cause-related activities through the businesses and their products. And in addition, the companies can apply the specialist knowledge of the NGO to create new innovations, products, services, and processes that are more sustainable.

Industries that are typically recognized as unsustainable in the public eye, such as the oil industry, the nuclear energy industry, and agriculture have taken a bigger responsibility for the pollution caused by their production and business. This has resulted in many actions to reduce CO_2 emissions and waste, to minimize risks in relation to environmental disasters, and attempts to reverse ambient images and perceptions of the companies, something more successful and effective than another. 'Greenwashing' is no longer an option with the transparency created by the internet and the social media. Thus, more sustainable solutions have to be sought in investments in development, innovation, and research.

At present there is an extensive focus on various sustainability standards, including global compact, ISO standards, responsible accounting practices, and CSR reporting, and which are well-established among particular larger, western companies. Yet, these standards will have to become mainstream for SMEs as well if they want to supply the large, CSR-based companies. The same standards can be expected to be integrated across companies in third world countries with the same argumentation. As a consequence, a harmonizing of sustainability and the level of social responsibility may develop globally. However, to follow these standards blindly and without relating them to what the company should gain from them and how they are embedded in the organization and among its global suppliers and value chain, will not generate the expected positive business impact and returns.

The fact that management and employee profiles must be able to think in sustainable business terms will also make demands on training modules offered, consultancy services, educational institutions, and HR departments in relation to talent management, recruitment, and training. Consequently, a number of questions emerge: What are the necessary skills for employees working with corporate sustainability and which courses will have to be offered in future at universities and other education providers to facilitate the right competences among students and employees? Do we need new types of managers and leaders? Does the company need to rethink the members of the board to ensure that the selected profiles have the ability to think sustainable business? And how should companies and their HR departments facilitate talent development to ensure that they have the right profiles with the right sustainability skills?

The employees and managers in today's and tomorrow's companies have to understand and comply with the requirements of sustainability. Thus, these questions should be asked and answered by both companies and their stakeholders, society, educational institutions as well as HR departments.

Companies will to an even greater extent experience more employees, particularly from Generation Y, making demands on corporate responsibility and sustainable management of employees, business partners, the environment, and society. In practice, this may implicate that employees in addition to their daily work and job description also demand CSR-related side jobs, such as volunteering in NGOs, so that they can live out their true passions and values. Work descriptions may also need to explain the company's fundamental values and the sustainable aspect of the job offered to a greater extent than before, as the new applicants or employees will want to evaluate whether their own values fit into those of the company.

The social media has open up the world and made many types of unsustainable behavior visible across the world. This also means that countries without democracy and liberty will be inspired, pressured and potentially attempt to pursue more social justice and environmental friendly solutions and also seek opportunities for democracy in sustainable societies in the Western world. Consequently, it may be expected that people from less democratic countries will increasingly seek employment in companies and countries that provide opportunities for personal and professional freedom. Yet, the virtual venues and channels for dialogue across continents have increased the opportunities for these transfers of knowledge and knowledge resources, also without people moving physically.

Finally, the next years will hopefully reveal more great examples of sustainable business and CSR that is better integrated throughout the organizations, their functions and value chains for all stakeholders and society to benefit from.

10

Concluding Remarks

CSR is here to stay. The global needs for sustainability and responsible behavior are greater now than ever before. The international companies and conglomerates are of such a size today that they visibly influence the social and environmental conditions and therefore play a central role in global sustainability. As the global, social, and environmental challenges do not seem to disappear, CSR will be a continuous premise for conducting business in a manner, which satisfies both the company, its stakeholders and society nationally and globally.

Sustainable business is increasingly becoming an integral part of organizations where CSR departments are being reorganized and getting integrated into other functions in ways that ensure the implementation of CSR strategies in business and the corporate strategy. At the same time, the nature and content of corporate sustainability evolves over time in correspondence to the growing requirements and knowledge of society and the company's stakeholders. This implies that what is expected of a sustainable business today will continuously change and grow so that what is 'nice-to-have' now will become 'need-to-have'.

In the future the boundaries between sustainable business in the company's value chain of various internal and external functions will disappear, and CSR practices across companies, industries, and different countries will be harmonized over time due the requirements of the international community emphasizing a common agenda for global sustainability. At the same time, customers and employees will play a more central role when it comes to define and prioritize corporate sustainability, especially in relation to sustainable development and corporate sustainable innovation. Also, the boundaries of public, private, state, government, and NGOs will get even more blurred as so many of the sustainable challenges of the world have to be solved across these boundaries if sustainable solutions are to be the final result.

The aim of this book is to bridge between theory and practice in shedding light on how sustainable business is defined, facilitated, and implemented across a company, its business, and functions (Communications and Sales, Production and Procurement, HRM and Administration, and R&D) and across industries and company sizes. Sustainability in theory is a key basis for the frame of reference and conceptual framework, which constitutes what we define as CSR today. However, CSR is acted out in practice, not in theory. There have been lots of great theoretical contributions on CSR, which are implemented in the explanation of sustainable business and how it is integrated. Thus, to put these theories into a context, several company cases have been presented to illustrate the many considerations to be made and the variety in which CSR can be integrated and sustainable business can be facilitated. It is therefore the hope of the author that this book will provide new knowledge, competences, and skills of sustainability for students and inspiration for employees and managers from both private and public companies, large and small, in supporting and facilitating more, sustainable businesses across the world.

Appendices

Appendix 1: The Fourteen Management Principles

The fourteen management principles by Henri Fayol were first published in the book publication, "*Administration Industrielle et Générale*" in 1917 and consists of the key factors, which according to Fayol's studies ensure successful management.

Principles of Management:

1. Division of work: This principle is the same as Adam Smith's "division of labor." Specialization increases output by making employees more efficient.
2. Authority: Managers must be able to give orders. Authority gives them this right. Note that responsibility arises wherever authority has been exercised.
3. Discipline: Employees must obey and respect the rules that govern the organization. Good discipline is the result of effective leadership, a clear understanding between management and workers regarding the organization's rules and the judicious use of penalties for infractions of the rules.
4. Unity of command: Every employee should receive orders from only one superior.
5. Unity of direction: Each group of organizational activities that have the same objective should be directed by one manager using one plan.
6. Subordination of individual interests to the general interest: The interests of any one employee or group of employees should not take precedence over the interests of the organization as a whole.
7. Remuneration: Workers must be paid a fair wage for their services.
8. Centralization: Centralization refers to the degree to which subordinates are involved in decision-making. Whether decision-making is centralized (to management) or decentralized (to subordinates) is a question of proper proportion. The task is to find the optimum degree of centralization for each situation.

9. Scalar chain: The line of authority from top management to the lowest ranks represents the scalar chain. Communications should follow this chain. However, if following the chain creates delays, cross-communications can be allowed if agreed to by all parties and superiors are kept informed.
10. Order: People and materials should be in the right place at the right time.
11. Equity: Managers should be kind and fair to their subordinates.
12. Stability of tenure of personnel: High employee turnover is inefficient. Management should provide orderly personnel planning and ensure that replacements are available to fill vacancies.
13. Initiative: Employees who are allowed to originate and carry out plans will exert high levels of effort.
14. Esprit de corps: Promoting team spirit will build harmony and unity within the organization.

Appendix 2: The Ten Principles of Global Compact

The UN Global Compact's ten principles are derived from the Universal Declaration of Human Rights, the International Labour Organization's Declaration on Fundamental Principles and Rights at Work, the Rio Declaration on Environment and Development, and the United Nations Convention Against Corruption. They are presented at UN's website and can be downloaded from the following link:
https://www.unglobalcompact.org/what-is-gc/mission/principles.

Human Rights

Principle 1: Businesses should support and respect the protection of internationally proclaimed human rights.
Principle 2: Make sure that they are not complicit in human rights abuses.

Labor

Principle 3: Businesses should uphold the freedom of association and the effective recognition of the right to collective bargaining.
Principle 4: The elimination of all forms of forced and compulsory labor.
Principle 5: The effective abolition of child labor; and
Principle 6: The elimination of discrimination in respect of employment and occupation

Environment

Principle 7: Businesses should support a precautionary approach to the following
Principle 8: Undertake initiatives to promote greater environmental responsibility; and
Principle 9: Encourage the development and diffusion of environmentally friendly technologies.

Anti-corruption

Principle 10: Businesses should work against corruption in all its forms, including extortion and bribery.

Appendix 3: A Selection of Global Websites on CSR and Sustainable Business

- Human Rights Compliance Assessment (HRCA) and the tool HRCA Quick Check:
  www.humanrightsbusiness.org
- Portal for Business Anti-corruption with handbooks and guidelines on anti corruption:
  www.business-anti-corruption.com
- Business Social Compliance Initiative is a joint European platform for retail textile and import companies, which provides knowledge on how to measure and improve the condition in the countries of foreign suppliers:
  www.bsci-cu.org
- UN's Global Compact is an initiative aimed at companies' social responsibility, which is based on international conventions on human rights, environment, anti-corruption. Global Compact consists of ten principles, which are described in Appendix 2:
  www.unglobalcompact.org
- Global Reporting Initiative (GRI) is a network-based organization that has developed guidelines and tools for reporting the documentation of the CSR in companies:
  www.globalreporting.org

Appendix 4: Selection of Tools for Integration of CSR and Sustainable Business

CSR:

- UN Global Compact
- Global Sullivan Principles of Corporate Social Responsibility
- Sigma Guiding Principles
- EFQM Framework for CSR
- Small Business Journey
- Ethos Business Social Responsibility Indicators
- Human Rights Compliance Assessment
- Business Principles for Countering Bribery

Customer activity:

- Fairtrade labels
- EU Ecolable

Employee activities

- Fair Labor Association Code of Conduct
- Amnesty International Human Rights Principles for Companies
- OHSAS 18001

Supplier activities

- CSR Compass
- Social Accountability 8000
- Ethical Trading Initiative

Stakeholders

- GRI Sustainability Reporting Guidelines
- AA1000 Framework
- AA1000 Assurance Standard
- Dow Jones Sustainability Indexes
- SME Key
- OECD Principles of Corporate Governance

Social and environmental activities

- London Benchmarking Group Model
- Community Mark

- ICC Business Charter for Sustainable Development
- ISO 14001
- Eco-Management and Audit Scheme (EMAS)
- Greenhouse Gas Protocol Initiative
- UNEP Resource Kit on Sustainable Consumption and Production

Source: Roepstorff and Serpa (2005).

References

Aagaard, A., & Lemmergaard, J. (2011). Effects of CSR integration in Danish smv's—Internally versus externally oriented CSR. Paper presented at Aarhus University CSR-workshop 24.11.2011.

Aagaard, A. (2011). *Idea and innovation management & leadership*. HansReitzels Forlag.

Aagaard, A. (2012). *CSR med succes—fra teori til praksis*. Gyldendal.

Aagaard, A., Eskerod, P., Hueman, M., & Ringhofer, C. (2016). Balancing management 'for' and 'of' stakeholders in projects and in temporary organizations Conference paper for EURAM 2016 Paris—Conference track SIG 10 Project Organizing.

Albareda, L., Lozano, J. M., Tencati, A., Midttun, A., & Perrini, F. (2008). The changing role of governments in corporate social responsibility: drivers and responses. *Business Ethics: A European Review, 17*(4), 347–363.

Albinger, H. S., & Freeman, S. J. (2000). Corporate social performance and attractiveness as an employer to different job seeking populations. *Journal of Business Ethics*, *28*(3), 243–53.

Al-Laham, A., Tzabbar, D., & Amburgey, T. L. (2011). The dynamics of knowledge stocks and knowledge flows: innovation consequences of recruitment and collaboration in biotech. *Industrial and Corporate Change*, *20*(2), 555–583.

American Psychological Association. (2015). Stress in America—paying with our health. Released February 4, 2015.

Amit, R., & Zott, G., (2001). Value creation in e-business. *Strategic Management Journal*, *22*(6–7), 493–520.

Angus-Leppan, T., Metcalf, L., & Benn, S. (2010). Leadership styles and CSR practice: An examination of sensemaking, institutional drivers and CSR leadership. *Journal of Business Ethics*, *93*, 189–213.

Argenti, P. A. (2004). Collaborating with activists: How starbucks works with NGOs, *California Management Review, 47*(1), 91–116.

Ashforth, B. E., & Gibbs, B. W. (1990). The double edge of organizational legitimation. *Organization Science*, *1*(2), 177–194.

Atkinson, A. A., Waterhouse, J. H., & Wells, R. B. (1997). A Stakeholder Approach to Strategic Performance Measurement. *Sloan Management Review, 38*(3), 25–37.

Austin, J. E. (2000). *The collaboration challenge: How Nonprofits and Businesses succeed through strategic alliances,* US: Jossey Bass Ltd.

Austin, J., Stevenson, H., & Wei-Skillern, J. (2006). Social and commercial entrepreneurship: Same, different, or both? *Entrepreneurship Theory and Practice, 30*(1), 1–22.

Austin. J. E. (2003). Marketing's role in cross-sector collaboration, *Journal of Nonprofit & Public Sector Marketing, 11*(1), 23–39.

Azapagic, A., Millington, A., & Collett, A. (2006). A methodology for integrating sustainability considerations into process design. *Chemical Engineering Research and Design, 84*(6), 439–452.

Baden-Fuller, C., & Morgan, M. S. (2010). Business models as models. *Long Range Planning* 43, 156–171.

Bansal, P. (2005). Evolving sustainability: A longitudinal study of corporate sustainable development, *Strategic Management Journal, 26*, 197–218.

Barnard, C. (1938). *The functions of the executives.*

Barrett, D. J. (2002). Change communication: using strategic employee communication to facilitate major change. *Corporate Communications: An International Journal, 7*(4), 219–231.

Basu, K., & Palazzo, G. (2008). Corporate social responsibility: A process model of sensemaking. *Academy of Management Review, 33*(1), 122–136.

Baumann, H., & Tillman, A. M. (2004). *The Hitch Hiker's Guide to LCA—An orientation in life cycle assessment methodology and application,* U.S.A., Studentlitteratur.

Baumgartner, R. J., & Ebner, D. (2010). Corporate sustainability strategies: Sustainability profiles and maturity levels. *Sustainable Development, 18*(2), 76–89.

Baumgartner, R. J. (2014). Managing corporate sustainability and CSR: A conceptual framework combining values, strategies and instruments contributing to sustainable development. *Corporate Social Responsibility & Environmental Management, 21*(5), 258–271.

Becker-Olsen, K. L., Cudmore, B. A., & Hill, R. P. (2006). The impact of perceived corporate social responsibility on consumer behavior. *Journal of Business Research, 59*(1), 46–53.

Beckman, T., Colwell, A., & Cunningham, P. H. (2009). The emergence of corporate social responsibility in Chile: the importance of authenticity and social networks. *Journal of Business Ethics, 86*, 191–206.

Beckmann, S. C., & Morsing, M. (2006). *Strategic CSR Communication.* DJØF Publication.

Beder, C. (2008). Regering vil tvinge virksomheder til at fortælle om samfundsansvar.

Beer, F., & Damgaard, B. (2007). Kommuner og Virksomheders samspil om socialt engagement. SFI—Det nationale forskningscenter for velfærd.

Bengtsson, L., Lakemond, N., Lazzarotti, V., Manzini, R., Pellegrini. L., & Tell, F. (2015). Open to a select few? Matching partners and knowledge content for open innovation performance. *Creativity and Innovation Management*, *24*(1), 72–86.

Bergman, B., & Klefsjö, B. (2010). *Quality from customer needs to customer satisfaction*, 3rd. Lund. Studentlitteratur.

Berlingske Tidende, 14. maj, s. 10.

Bhattacharya, C. B., & Sen, S. (2004). When, why, and how consumers respond to social initiatives. *California Management Review, 47*(1), 9–24.

Bhattacharya, C. B., Sen, S., & Korschun, D. (2008). Using corporate social responsibility to win the war for talent. (cover story). *MIT Sloan Management Review*, *49*(2), 37–44.

Bird, R., Hall, A. D., Momentè, F., & Reggiani, F. (2007). What corporate social responsibility activities are valued by the marked? *Journal of Business Ethics, 76*(2), 189–206.

Birkin, F., Cashman, A., Koh, S. C. L., & Liu, Z. (2009a). New sustainable business models in China. *Business Strategy and the Environment* 18, 64–77.

Birkin, F., Polesie, T., & Lewis, L. (2009b). A new business model for sustainable development: an exploratory study using the theory of constraints in Nordic organizations. *Business Strategy and the Environment, 18*, 277–290.

Bisgaard, T. (2009). CSI Corporate Social Innovation—Companies participation in solving global challenges. Erhvervs- & Byggestyrelsen.

Bloom, P. N., & Chatterji, A. K. (2009). Scaling social entrepreneurial impact. *California Management Review*, *51*(3), 114–133.

Blowfield, M., & Frynas, J. (2005). Setting new agenda: critical perspectives on corporate social responsibility, *International Affairs*, *81*(3), 499–513.

Blowfield, M. & Murray, A. (2008), Corporate Responsibility—a critical introduction. Oxford University Press, UK.

Boettke, P., & Coyne, C. (2008). The Political Economy of the Philanthropic Enterprise. In G. Shockley, P. Frank, & R. Stough, (Eds.), *Non-market Entrepreneurship: Interdisciplinary Approaches*, (pp. 71–88). Edward Elgar Publishing.

Boks, C., & Tempelman, E. (1998). Future disassembly and recycling technology: Results of a Delphi study, *Futures, 30*(5), 425–442.

Boons, F., & Lüdeke-Freund, F. (2013). Business models for sustainable innovation: state-of-the-art and steps towards a research agenda. *Journal of Cleaner Production*, *45*, 9–19.

Boons, F., & Mendoza, A. (2010). Constructing sustainable palm oil: how actors define sustainability. *Journal of Cleaner Production, 18*, 1686–1695.

Boons, F. (2009). *Creating ecological value.* In: *An evolutionary approach to business strategies and the natural environment.* Elgar, Cheltenham.

Boons, F., Montalva, C., Quist, J., & Wagner, M. (2013). Sustainable innovation, business models and economic performance: an overview. *Journal of Cleaner Production*, *45*, 1–8.

Bos-Brouwers. H. E. J. (2010). Corporate sustainability and innovation in SMEs: Evidence of themes and activities in practice. *Business Strategy and the Environment, 19*(7), 417–435.

Bowen, F. E, Cousins, P. D, Lamming, R. C., & Farukt, A. C. (2001). The role of supply management capabilities in green supply. *Production and Operations Management, 10*(2), 174–189.

Bowen, D. E., & Ostroff, C. (2004). Understanding HRM-firm performance linkages: The role of the "strength" of the HRM System. *Academy of Management Review*, *29*(2), 203–221.

Bradbury-Huang, H. (2010). Sustainability by collaboration: The Seer case, *Organizational Dynamics*, *39*, 335–344.

Brammer, S., & Pavellin, S. (2004). Voluntary social disclosures by large UK companies. Business Ethics: *A European Review*, *13*(2/3), 86–99.

Brousseau, J., Chiagouris, L., & Fernandez Brusseau, R. (2013). Corporate social responsibility: to yourself be true. *Journal of Global Business and Technology*, *9*(1), 53–63.

Buono, A. F. et al. (2009). Organizational change: Building organizational change capacity. Academy of Management, Conference Paper, Illionis, 08-2009.

Burchell, J., & Cook, J. (2013). Sleeping with the Enemy? Strategic transformations in business–NGO relationships through stakeholder dialogue, *Journal of Business Ethics*, *113*(3), 505–518.

Burke, L., & Logsdon, J. M. (1996). Corporate social responsibility, pays off, *Long Range Planning*, *29*(4), 437–596.

Business for Social Responsibility. (2000). Introduction to corporate social responsibility. http://www.khbo.be/~lodew/cursussen/4eingenieurCL/The%20Global%20Business%20Responsibility%20Resource%20Center.doc

Cannon, T. (2012). *Corporate responsibility: governance, compliance, and ethics in a sustainable environment,* (2nd Edn.). New York, NY: Pearson.

Canter, M. R. (1999). From spare change to real change: The social sector as beta site for business innovation. Harvard Business Review.

Carbone, V., Moatti, V., & Vinzi, V. E. (2012). Mapping corporate responsibility and sustainable supply chains: an exploratory perspective. *Business Strategy and the Environment, 21*(7), 475–494.

Carroll, A. B. (1979). A three-dimensional conceptual model of corporate social performance. *Academy of Management Review*, *4*, 497–505.

Carroll, A. B. (1991). The pyramid of corporate social responsibility: Toward the moral management of organizational stakeholders. *Business Horizons*, July/August, Elsevier.

Carroll, A. B. (1999). Corporate social responsibility. *Business and Society*, *38*(3), 268–296.

Carroll, A. B. & Buchholtz, A. K. (2006), Business and society. ethics and stakeholder management, Mason, Ohio, Thomson/South-Western.

Carroll, A. B., & Shabana, K. M. (2010). The business case for corporate social responsibility: A review of concepts, research and practice. *International Journal of Management Reviews*. DOI: 10.1111/j.1468-2370.2009.00275.x

Carter, C. R., &. Jennings, M. M. (2002a). Social responsibility and supply chain relationships. *Transportation Research, Part E 38E*(1), 37–52.

Carter, C. R., & Jennings, M. M (2002b). Logistics social responsibility: an integrative framework. *Journal of Business Logistics, 23*(1), 145–180.

Carter, C. R., & Jennings, M. M. (2004). The role of purchasing in corporate social responsibility: a structural equation analysis. *Journal of Business Logistics, 25*(1), 145–186.

Casadesus-Masanell, R., & Ricart, J. E. (2011). How to design a winning business model. *Harvard Business Review*, *89*, 100–107.

Castka, P., & Balzarova, M. A. (2007). A critical look on quality through CSR lenses: Key challenges stemming from the development of ISO 26000. *International Journal of Quality & Reliability Management, 24*(7), 738–752.

Castka, P., Balzarova, A., & Bamber, C. (2004). How can SMEs effectively implement the CSR agenda? A UK case study perspective. *Corporate Social Responsibility and Environmental Management, 11*(3), 140–149.

Chandler, A. (1962) *Strategy and Structure.* MIT Press: Cambridge, MA.

Charter, M., & Clark, T. (2007). Sustainable innovation—key conclusions from sustainable innovation conferences 2003–2006. The Centre for Sustainable Design. University College for the Creative Arts. May 2007. www.cfsd.org.uk

Chatterjee, S. (2013). Simple Rules for Designing Business Models. *California Management Review*, *55*, 97–124.

Chesbrough, H., & Brunswicker. S. (2014). A fad or a phenomenon? The adoption of open innovation practices in large firms. *Research-Technology Management*, March–April, 16–25.

Chesbrough, H. (2007). Business model innovation: It is not just about technology anymore. *Strategy and Leadership*, *35*(6), 12–17.

Chesbrough, H. (2010). Business model innovation: Opportunities and barriers. *Long Range Planning*, *43*, 354–363.

Chesbrough, H., & Rosenbloom, R., (2002). The role of the business model in capturing value from innovation. *Industrial and Corporate Change*, *11*(3), 529–556.

Chesbrough, H. W. (2005). *Open Innovation: The New Imperative for Creating And Profiting from Technology.* Harvard Business Review Press.

Chesbrough, H., Ahern, S., Finn, M., & Guerraz, S. (2006). Business models for technology in the developing world: the role of non-governmental organizations. *California Management Review*, *48*, 48–61.

Chesbrough, H., Vanhaverbeke, W., & West, J. (2006). *Open Innovation: Researching a New Paradigm.* Oxford: Oxford University Press.

Chiodo, J. D., & Boks, C. (2002). Assessment of end-of-life strategies with active disassembly using smart materials. *The Journal of Sustainable Product Design*, *2*(1–2), 69–82.

Choi, J., Nies, L., & Ramani, K. (2008). A framework for the integration of environmental and business aspects toward sustainable product development. *Journal of Engineering Design*, *19*(5), 431–446.

Christensen, C. M. (1997). *The Innovators Dilemma: When new technologies cause great firms to fail,* Harvard Business School Press, Boston, MA.

Christensen, G. M., & Raynor, M., (2003). *The Innovator's Solution.* Boston: Harvard Business School Press.

Christensen, L. T., M. Morsing & O. Thyssen (2011). The polyphony of corporate social responsibility. Deconstructing accountability and transparency in the context of identity and hypocrisy, In: Cheney, G, May, S. & Munshi, D. (Eds.), Handbook of Communication Ethics, pp. 457–474). New York: Lawrence Erlbaum Publishers.

Ciliberti, F., Pontrandolfo, P., & Scozzi, B. (2008). Investigating corporate social responsibility in supply chains: A SME perspective. *Journal of Cleaner Production, 16*(15), 1579–1588.

Cohen, E. (2010). *CSR for HR: A Necessary Partnership for Advancing Responsible Business Practices.* Greenleaf Publishing.

Cohen, E. (2010). *CSR for HR,* Greenleaf, Sheffield.

Copenhagen Business School (CBS) (2009). HR og CSR—hvordan spiller det sammen? FBE—Forum for Business Education, June 2009.

Corbett, C. J., & Klassen, R. D. (2006). Extending the horizons: environmental excellence as key to improving operations. *Manufacturing & Service Operations Management, 8*(1), 5–22.

Cordano, M., & Frieze, I. H. (2000). Pollution reduction preferences of U.S. environmental managers: applying Ajzen's theory of planned behavior. *Academy of Management Journal, 43*(4), 627–641.

Cornelissen, J. (2004). Corporate communications: theory and practice, Sage: London.

Corporate Register. (2011). CSR report 2011. www.corporateregister.com

Cramer, J. (2005). Experiences with structuring corporate social responsibility in Dutch industry. *Journal of Cleaner Production, 13*(6), 583–592.

Cramer, J., van der Heijden, A., & Jonker, J. (2006). Corporate Social Responsibility: Making SenseThrough Thinking and Acting. *Business Ethics: A European Review, 15*(4), 380–389.

Crane, A., McWilliams, A., Matten, D., Moon, J., & Siegel, D. S. (2008). *The Oxford Handbook of Corporate Social Responsibility.* UK, Oxford University Press.

Crawford, D., & Scaletta, T. (2005). The balanced scorecard and corporate social responsibility: Aligning values for profit. *CMA Magazine*, October 23, 2005.

Crook, C. (2005). The good company. *Economist, 37*, January, 31–41.

Dahan, N. M., Doh, J. P., Oetzel, J., & Yaziji, M. (2010). Corporate-NGO collaboration: co-creating new business models for developing markets. *Long Range Planning, 43*, 326–342.

Dahlander, L., & Gann, D. M. (2010) How open is innovation? *Research Policy*, *39*, 699–709.

Dahlsrud, A. (2008). How corporate social responsibility is defined:an analysis of 37 Definitions. *Corporate Social Responsibility and Environmental Management, 15*, 1–13.

Daly, H. E. (1990). Toward some operational principles of sustainable development. *Ecological Economics, 2*(1), 1–6.

Darnall, N., Henriques, I., & Sadorsky, P. (2008). Do environmental management systems improve business performance in an international setting? *Journal of International Management*, *14*, 364–376.

Datta, D. K., Guthrie, J. P., Basuil, D., & Pandey, A. (2010). Causes and effects of employee downsizing: A review and synthesis. *Journal of Management*, *36*(1), 281–348.

David, F. (1989). *Strategic Management*. Merrill Publishing Company: Columbus.

Dawkins, C., & Ngunjiri, F. W. (2008). Corporate social responsibility reporting in South Africa. *Journal of Business Communication*, *45*(3), 286–307.

Dawkins, J. (2005). Corporate responsibility: the communication challenge. *Journal of Communication Management*, *9*(2), 108–119.

De Bakker, F. G. A., Groenewegen, P., & Den Hond, F. (2005). A bibliometric analysis of 30 Years of research and theory on corporate social responsibility and corporate social performance. *Business & Society*, *44*(3), 283–317.

De Saa-Perez, P., & Garcia-Falcon, J. M. (2002). A resource-based view of human resource management and organisational capabilities development. *International Journal of Human Resource Management, 13*, 123–140.

De Sousa, M. C. (2006). The sustainable innovation engine, *The Journal of Information and Knowledge Systems*, *34*(6), 398–405.

De Winne, S., & Sels, L. (2010) Inter relationships between human capital, HRM and innovation in Belgian start-ups aiming at an innovation strategy. *The International Journal of Human Resource Management, 21*(11), 1863–1883.

Deloitte. (2011). *CSR forankringi danske virksomheder*. Report.

DI (2007a). "Sæt fokus på kulturen." Presentation on Conference om LEAN in service og administration.

DI (2007b). Produktivitetsundersøgelse 2007. "Brugen og nytten af Lean. Analyse på basis af en spørgeskemaundersøgelse blandt 500 ledere." Report.

Di Domenico, M. L., Haugh, H., & Tracey, P. (2010). Social bricolage: theorizing social value creation in social enterprises. *Entrepreneurship Theory and Practice, 34*(4), 681–703.

DI Handel. (2008). Handel og ansvar—sådan arbejderdin handelsvirksomhed med samfundsansvar ogbæredygtighed. Report.

DiBella, A., & Nevis, E. (1998). *How organizations learn: An integrated strategy for building learning capability.* Jossey-Bass Press.

Diller, J. (1999). A social conscience in the global marketplace? Labour dimensions of codes of conduct, Social labelling and investor initiatives. *International Labour Review, 138*(2), 99–129.

Doganova, L., & Eyquem-Renault, M. (2009). What do business models do? Innovation devices in technology entrepreneurship. *Research Policy, 38*, 1559–1570.

Doh, J. P., & Guay, T. R. (2006). Corporate social responsibility, Public Policy, and NGO Activism in Europe and the United States: An institutional-stakeholder perspective. *Journal of Management Studies, 43*(1) 47–73.

Doh, J. P., & Guay, T. R. (2004). Globalization and Corporate Social Responsibility: How Non-Governmental Organizations influence labor and environmental codes of conduct, *Management International Review, 44*(2).

Dortmans, P. J. (2005). Forecasting, backcasting, migration landscapes and strategic planning maps. *Futures, 37*(4), 273–285.

Dreyer, L., et al. (2006). A framework for social life cycle impact assessment. *The International Journal of Life Cycle Assessment, 11*(2), 88–97.

Dreyer, R. (2011). Indkøbere ved for lidt om bæredygtige indkøb. Article publiced on May 4th, 2011 at www.csr.dk/leverandor

Dryzek, J. S., (2005). *The Politics of the Earth: Environmental Discourses.* Oxford University Press.

Dutot, V., Galvez, E. L., & Versailles, D. W. (2016). CSR communications strategies through social media and influence on e-reputation. *Management Decision, 54*(2), 363–389.

Dyllick, T., & Hockerts, K. (2002). Beyond the business case for corporate sustainability. *Business Strategy and the Environment, 11*, 130–141.

Ehnert, I. (2006). Sustainability Issues in Human Resource Management: Linkages, Theoretical Approaches, and Outlines for an Emerging Field. Paper presented at 21st EIASM Workshop on SHRM, March 30–31, 2006, Birmingham.

Ehnert, I. (2009). Sustainable human resource management: a conceptual and exploratory analysis from a paradox perspective. Contributions tomanagement science series. Heidelberg: Physica/Springer.

Elkinton, J., (1997). *Cannibals with forks: Triple Bottom Line of 21st Century Business*, Capstone Publisher Limited, Oxford.

Elliott, K. A., & Freeman, R. B. (2000). White Hats or DonQuixotes: Human Rights Vigilantes in the Global Economy. National Bureau of Economic Research.

Ellis, T. (2010). *The New Pioneers*, Wiley.

Erdem, T., & Swait, J. (2004). Brand credibility, brand consideration, and choice. *Journal of Consumer Research, 31*(1), 191–198.

Eriksen, M., Fischer, T., & Mønsted, L. (2005). *God leanledelse i administration og service*. Børsens Forlag.

Esping-Andersen, G. (1990). *The three worlds of welfare capitalism*. Polity Press. Blackwell Publishing Ltd.

EU Commission. (2001). Green Paper. *Promoting a European framework for Corporate Social Responsibility.*

EU Commission. (2003). Green Paper. *Entrepreneurship in Europe.*

Faber, N., Jorna, R. & Van Engelen, J., (2005). The sustainability of "sustainability"—A study into the conceptual foundations of the notion of "sustainability", *Journal of Environmental Assessment Policy and Management*, *7*(1), 1–33.

Fayol, H. (1916). *Administration industrielle et générale; prévoyance, organisation, commandement, coordination, controle*, H. Dunod & E. Pinat.

Ferradás, P. G., & Salonitis, K. (2013). Improving Changeover Time: A Tailored SMED Approach for Welding Cells. *Procedia CIRP*, 7, 598–603. Forty Sixth CIRP Conference on Manufacturing Systems 2013.

Fischer, F. (2000). *Citizens, experts, and the environment: The politics of local knowledge*. Durham: Duke University Press.

Fombrun, C. J. (2005). The Leadership Challenge: Building Resilient Corporate Reputations. In J. P. Doh and S. A. Stumpf (Eds.). *Handbook on Responsible Leadership and Governance in Global Business. Edward* Elgar, Cheltenham, pp. 54–68.

Forrester Research. (2008). Global Enterprise Web 2.0 Market Forecast: 2007 to 2013. www.forrester.com.

Fox, A. (2007). Corporate social responsibility pays off. *HR Magazine, 52*(8), 43–47.

Freeman, R. E., Harrison, J. S., & Wicks, A. C. (2007). *Managing for Stakeholders: Survival, Reputation, and Success.* Yale University Press.

Freeman, R. E., Harrison, J. S., Wicks, A. C., Parmar, B. L., & De Colle, S. (2010). Stakeholder Theory: The State of the Art. Cambridge: Cambridge University Press.

Freeman, R. E. (1984). *Strategic management: A stakeholder approach.* Pitman.

Frenkel, S. J. (2001). Globalization, athletic footwear commodity chains and employment relations in china. *Organization Studies 22*(4), 531–562.

Friedman, A. L., & Miles, S. (2002). Developing Stakeholder Theory. *Journal of Management Studies, 39*(1), 1–21.

Friedman, M. (1970). The social responsibility of business is to increase its profits. *The New York Times*, (September), 33.

Gabor, A., & Mahoney, J. T. (2010). Chester Barnard and the Systems Approach to Nurturing Organizations. Working paper.

Galbreath, J. (2006). Corporate social responsibility strategy: Strategic options, global considerations. *Corporate Governance*, *6*(2), 175–187.

Gallopin, G. (1997). Indicators and their use: information for decision-making. In: B. Moldan, & S. Billharz (Ed.) *Sustainability indicators: Report of the project on indicators of sustainable development.* John Wiley and Sons Ltd.

Garcia-Marza, D. (2005). Trust and dialogue: Theoretical approached to ethics auditing. *Journal of Business Ethics*, *57*, 209–219.

Garvare, R., & Isaksson, R. (2001). Sustainable development: Extending the scope of business excellence models. *Measuring Business Excellence*, *5*(3), 11–15.

Gehani, R. R. (2002). Chester Barnard's "executive" and the knowledge-based firm. *Management Decision, 40*(10), 980–991.

Gehin, A., Zwolinski, P., & Brissaud, D. (2008). A tool to implement sustainable end-of-life strategies in the product development phase. *Journal of Cleaner Production*, *16*(5), 566–576.

Gendron, C. (2009). ISO 26000: Towards a Social Definition of Corporate Social Responsibility. ISA Annual Convention 2009, Conference paper, 1–13.

Giberson, T. R., Resick, C. J., Dickson, M. W., Mitchelson, J. K., Randall, K. R., Clark, M. A. (2009). Leadership and organizational culture:

linking CEO characteristics to cultural values. *Journal of Business and Psychology, 24*(2), 123–137.
Giddens, A. (2003). *Runaway world.* New York: Routledge.
Glade, B. (2008). Human resources: CSR and business sustainability—HR's leadership role. *New Zealand Management, 55*(9), 51–52.
Gond, J. P., Igalens, J., Swaen, V., & Akremi, A. E. (2011). The Human Resources Contribution to Responsible Leadership: An Exploration of the CSR-HR Interface. *Journal of Business Ethics, 98*, 115–132.
Gond, J., Igalens, J., Sswaen, V., & Akremi, A. (2011). The human resources contribution to responsible leadership: An exploration of the CSR-HR interface. *Science & Business Media, 98*, 115–132.
Goodland, R. (1995). The concept of environmental sustainability. *Annual Review of Ecology and Systematics,* 26, 1–24.
Govindarajan, V., & Trimble, G. (2005). *Ten rules for strategic innovators: From idea to execution.* Boston: Harvard Business School Press.
Grayson, D. (2004). How CSR Contribute to the Competitiveness of Europe in a More Sustainable World. The WorldBank Institute and the CSR Resource Centre, Netherlands.
GRI. (2015). Leading for a New Era of Sustainability. GRI's Combined Report 2014–2015.
Gupta, M. (1995). Environmental management and its impact on the operations function. *International Journal of Operations & Production Management, 15*(8), 34–51.
Gürtler, H. (2009). Etisk handel er også et offentligt anliggende. Artikel fra d. 7. september 2009.
Hansen, E. G., Große-Dunker, F., Reichwald, R. (2009). Sustainability innovation Cube—a framework to evaluate sustainability-oriented innovations. *International Journal of Innovation Management, 13*, 683–713.
Hanssen, O. (1999). Sustainable product systems—experiences based on case projects in sustainable product development. *Journal of Cleaner Production, 7*(1), 27–41.
Hart, S. L., & Milstein, M. B. (1999). Global sustainability and the creative destruction of industries. *Sloan Management Review, 41*, 23–33.
Harvard University and Foundation Strategy Group. (2005). Competitive social responsibility: Uncovering the Economic Rationale for Corporate Social Responsibility among Danish Small- and Medium-Sized Enterprises. Report.
Heck, S., Rogers, M., & Carroll, P. (2014). *Resource Revolution: How to Capture the Biggest Business Opportunity in a Century.* Amazon Publishing.

Hemingway, C. A., & Maclagan, P. W. (2004). Managers' Personal Values as Drivers of Corporate Social Responsibility. *Journal of Business Ethics*, *50*(1), 33–44.

Hemingway, C. A. (2005). Personal values as a catalyst for corporate social entrepreneurship. *Journal of Business Ethics*, *60*(3), 233–249.

Hernáez, O., Zugasti, I., Waltersdorfer, G., Matev, D., Assenova, M., Jonkute, G., Staniskis, J., Schoenfelder, T., Bogataj, M., Møller, J. D., Hirsbak, S., Schmidt, K., Christiansen, K., Fondevila, M., & Aranda, J. (2012). Corporate Social Responsibility on SMEs, Paper presented at 15th European Roundtable on Sustainable Consumption and Production, Bregenz, Østrig.

Hervani, A., Helms, M. M., & Sarkis, J. (2005). Performance measurement for green supply chain management. *Benchmarking, 12*(4), 330–353.

Hess, D. (1999). Social reporting: A reflexive law approach to corporate social responsiveness. *Journal of Corporate Law*, *Fall, 25*(1), 41–48.

Highhouse, S., Brooks, M. E., & Gregarus, G. (2009). An organizational impression management perspective on the formation of corporate reputations. *Journal of Management, 35*(6), 1481–1493.

Hildebrand, D., Sen, S., & Bhattacharya, C. (2011). Corporate social responsibility: A corporate marketing perspective. *European Journal of Marketing, 45*(9/10), 1353–1364.

Hildebrandt, S. & Brandi, S. (2005). Ledelse af forandring. L&R Business.

Hill, R. P., Ainscough, T., Shank, T., & Manullang, D. (2007). Corporate social responsibility and socially responsible investing: A global perspective. *Journal of Business Ethics, 70*, 165–174.

Hoeffler, S., Bloom, P., & Keller, K. L. (2010). Understanding stakeholder responses to corporate citizenship initiatives: Managerial guidelines and research directions. *Journal of Public Policy and Marketing, 29*(1), 78–88.

Hohnen, P. (2007). *Corporate social responsibility, An implementation guide for business.* International Institute for Sustainable Development, Manitoba, Canada.

Holling, C. S. (2001). Understanding the Complexity of Economic, Ecological, and Social Systems. *Ecosystems, 4*, 390–405.

Holmberg, J., & Robèrt, K. H. (2000). Backcasting—a framework for strategic planning. *International Journal of Sustainable Development and World Ecology* 7: 291–308.

Hongwei, H., & Yan, L. (2011). CSR and service brand: the mediating effect of brand identification and moderating effect of service quality. *Journal of Business Ethics*; *100*(4), 673–688.

Hopkins, M. (2005). Measurement of Corporate Social Responsibility. *International Journal of Management and Decision Making, 6*(3/4).

Hudson, S., & Roloff, J. (2010). In search of sustainability? SMEs in Britanny, France. In Spence, L., & Painter-Morland, M. (Eds.) Ethics in small and medium sized enterprises: a global commentary. Dordrecht, Netherlands: Springer.

Huizingh, E. K. R. E. (2011) Open innovation: State-of-the-art and future perspectives. *Technovation*, *31*, 2–9.

Hull, C. E., & Rothenberg, S. (2008). Firm performance: the interactions of corporate social performance with innovation and industry differentiation. *Strategic Management Journal*, *29*(7), 781–789.

Husted, B. W. (2005). Risk management, real options, and corporate social responsibility. *Journal of Business Ethics, 60*, 175–183.

Husted, B. W. (2000). A contingency theory of corporate social responsibility, *Business & Society, 39*(1), 24–48.

Ilomaki, M., & Melanen, M. (2001). Waste minimisation in small and medium-sized enterprises—do environmental management systems help? *Journal of Cleaner Production*, *9*, 209–217.

Inyang, B., Awa, H., & Enuoh, R. (2011). CSR-HRM nexus: Defining the role engagement of the human resources professionals. *International Journal of Business and Social Science*, *2*(5), 118–126.

Isaksson, R. (2005). Economic sustainability and the cost of poor quality. *Corporate Social Responsibility and Environmental Management*, *12*(4), 197–209.

Isaksson, R. (2006). Total quality management for sustainable development: Process based system models. *Business Process Management Journal*, *12*(5), 632–645.

Iverson, R. D., & Zatzick, C. D. (2011). The effects of downsizing on labor productivity: The value of showing consideration for employees' morale and welfare in high-performance work systems. *Human Resource Management*, *50*(1), 29–44.

James, P. (1997). The sustainable circle: A new tool for product development and design. *Journal of Sustainable Product Design*, *2*, 52–57.

Jenkins, H. (2006). Small Business Champions for Corporate Social Responsibility. *Journal of Business Ethics, 6*(3), 241–256.

Jenkins, R. (2005). Globalization, Corporate Social Responsibility and Poverty. *International Affairs, 81*, 525–540.

Johansson, G. (2002). Success factors for integration of ecodesign in product development: a review of state of the art. *Environmental Management and Health*, *13*(1), 98–107.

Jones, B. (2009). Corporate reputation in the era of Web 2.0: the case of primark. *Journal of Marketing Management*, *25*(9), 927–939.

Jones, D. A., Willness, C. R., & Madey, S. (2010). Why are job seekers attracted to socially responsible companies? Testing underlying mechanisms. *Academy of Management Proceedings*, *2010*(1), 1–6.

Juel, K., Sørensen, J., & Brønnum-Hansen, H. (2006). Psykisk arbejdsbelastning. In K. Juel, J. Sørensen, & H. Brønnum-Hansen (Eds). *Risikofaktorer og folkesundhed i Danmark* (pp. 247–263). Statens Institut for Folkesundhed.

Juholin, E. (2004). For business or the good of all? A Finnish approach to corporate social responsibility. *Corporate Governance*, *4*(3), 20–31.

Julian, S. D., Ofori-Dankwa, J. C., & Justis, R. T. (2008). Understanding strategic responses to interest group pressures. *Strategic Management Journal*, *29*, 963–984.

Kaebernick, H., Kara, S., & Sun, M. (2003). Sustainable product development and manufacturing by considering environmental requirements. *Robotics and Computer-Integrated Manufacturing*, *19*(6), 461–468.

Kanter, R. M. (1999). From spare change to real change: the social sector as a beta site for business innovation. *Harvard Business Review*, *77*, 123–132.

Kharbili, M. E., Stein, S., Markovic, I., & Pulvermüller, E. (2008). Towards a framework for semantic business process compliance management. *Proceedings of GRCIS*, 2008.

Kim, C. W., & Mauborgne, R. (2005). *Blue Ocean Strategy*. Harvard Business School Press.

Kiron, D., Kruschwitz, N., Haanaes, K., Reeves, M., Fuisz-Kehrbach, S. K., & Kell, G. (2015). Joining forces: collaboration and leadership for sustainability. *MIT Sloan Management Review*, Research report January, available at: http://sloanreview.mit.edu/projects/joining-forces/ (accessed 25 May, 2016).

Kistruck, G. M., & Beamish, P. W. (2010). The interplay of form, structure and embeddedness in social intrapreneurship. *Entrepreneurship Theory and Practice*, *34*(2), 735–761.

Klay, W. E. (2015). The enlightenment underpinnings of the public procurement profession. *Journal of Public Procurement*, *15*(4), 439–457.

Klefsjö, B., Bergquist, B., & Garvare, R. (2008). Quality management and business excellence, customers and stakeholders: Do we agree on what we are talking about, and does it matter? *The TQM Journal, 20*(2), 120–129.

Kleindorfer, P. R., Singhal, K., & Wassenhove, L. N. (2005). Sustainable operations management. *Production and Operations Management, 14*(4), 482–492.

Klöpffer, W. (2003). Life-cycle based methods for sustainable product development. *The International Journal of Life Cycle Assessment, 8*(3), 157–159.

Kotter, J. (1990). *A force for change: how leadership differs from management.* The Free Press.

Kourula, A., & Halme, M. (2008). Types of corporate responsibility and engagement with NGOs: an exploration of business and societal outcomes. *Corporate Governance, 8*(4), 557–570.

Kovacs, R. (2006). Interdisciplinary bar for the public interest: What CSR and NGO frameworks contribute to the public relations of British and European activists. *Public Relations Review, 32*(4), 429–431.

KPMG. (2013a). Carrots and Sticks—Sustainability reporting policies worldwide—today's best practice, tomorrow's trends. Report.

KPMG. (2013b), The KPMG survey of Corporate Responsibility Reporting 2013. KPMG, Amsterdam, available at: www.kpmg.nl (accessed 26 May 2016).

Krier, J. M. (2007). Fair Trade: New Facts and Figures from an Ongoing Success Story. A Report on Fair Trade in 33 Consumer Countries. Fair Trade Advocacy Office: Brussels.

Kuhn, T., & Deetz, S. (2008). Critical theory and corporate social responsibility—can/should we get beyond cynical reasoning? In Crane et al. (Ed.) *The Oxford Handbook of Corporate Social Responsibility*, (pp. 173–196). UK: Oxford University Press.

Lakshman, C., Ramaswami, A., Alas, R., Kabongo, J., & Rajendran, P. (2014). Ethics Trumps Culture? A Cross-National Study of Business Leader Responsibility for Downsizing and CSR Perceptions. *Journal of Business Ethics, 125*(1), 101–119.

Lam, H., & Khare, A. (2010). HR's crucial role for successful CSR. *Journal of International Business Ethics, 3*(2), 3–82.

Larson, A. L. (2000). Sustainable innovation through an entrepreneurship lens, *Business and the Environment, 9*, 304–317.

Lawrence, P., & Lorsch, J. (1967). *Organization and Environment.* Harvard Business School Press, Boston, MA.

Leigh, J., & Waddock, S. (2006). The emergence of total responsibility management systems: J Sainsbury's (plc) Voluntary Responsibility management systems for global food retail supply chains. *Business and Society Review, 111*(4), 409–426.

Leuenberger, D. Z., & Wakin, M. (2007). Sustainable Development in Public Administration Planning: An exploration of social justice, equity, and citizen inclusion. *Administrative Theory & Praxis, 29*(3), 394–411.

Ligeti, G., & Oravecz, A. (2009). CSR communication of corporate enterprises in Hungary. *Journal of Business Ethics, 84*(2), 137–149.

Lindgren, P, Aagaard, A., & Ulldahl, L. (2015). How to establish knowledge sharing from the very first moment in critical and risky Business Model Innovation project. IFKAD conference proceedings 2015, Bari, Italy.

Lindgren, P., & Aagaard, A. (2014). The sensing business model. *Wireless Personal Communication, 76*, 291–309.

Lindgren, P., (2013). The Business Model Cube. *Journal of Multi Business Model Innovation and Technology, 1*(2). River Publisher.

Linnenluecke, M. K., Russell, S. V., & Griffiths, A. (2007). Subcultures and sustainability practices: the impact on understanding corporate sustainability. *Business Strategy and the Environment.* Online Journal.

Linton, J. (1999). Electronic products at their end-of-life: options and obstacles. *Journal of Electronics Manufacturing, 9*(1), 29–40.

Lis, B. (2012). The relevance of corporate social responsibility for a sustainable human resource management: An analysis of organizational attractiveness as a determinant in employees' selection of a (potential) employer. *Management Review, 23*(3), 279–295.

Ljungberg, L. Y. (2007). Materials selection and design for development of sustainable products. *Materials & Design, 28*(2), 466–479.

Lodsgård, L., & Aagaard, A. (2016). The four archetypes of Business-NGO collaborations in creating sustainable innovation. Conference paper. 23rd IPDMC Innovation & Product Development Management Conference, Glasgow, UK.

Longo, M., Mura, M., & Bonoli, A. (2005). Corporate social responsibility and corporate performance: the case of Italian SMEs. *Corporate Governance, 5*, 28–43.

Loorbach, D., van Bakel, J. C., Whiteman, G., & Rotmans, J. (2009). Business strategies for transitions towards sustainable systems, *Business strategy and the environment.*

Lopez, M. V., Garcia, A., & Rodriguez. L. (2007). Sustainable development and corporate performance: A study based on the Dow Jones Sustainability Index. *Journal of Business Ethics, 75*(3), 285–300.

Louche, C. (2010). *Innovative CSR: From Risk Management to Value Creation.* Greenleaf Publishing.

Loureiro, M. L, & Lotade, J. (2005). Do fair trade and eco-labels in coffee wake up the consumer conscience? *Ecological Economics, 53*(1), 129–138.

Lovins, A. B., Lovins, L. H., Hawken, P. (1999). A road map for natural capitalism. *Harvard Business Review*, 1–14 (HBR paperback reprint 2000).

Lüdeke-Freund, F., 2009. Business Model Concepts in Corporate Sustainability Contexts. From rhetoric to a generic template for 'business models for sustainability'. Centre for Sustainability Management, Lüneburg.

Lyon, T. P., & Montgomery, A. W. (2013). Tweetjacked: The impact of social media on corporate greenwash. *Journal of Business Ethics, 118*(4), 747–757.

Magretta, J. (2002). Why Business Models Matter. *Harvard Business Review, 80*, 86–92.

Mahoney, J. T. (2002). The relevance of Chester I. Barnard's teachings to contemporary management education: Communicating the aesthetics of management. *International Journal of Organization Theory and Behavior, 5*, 159–172.

Maignan, I., & Ferrell, O. C. (2004). Corporate social responsibility and marketing: An integrative framework. *Journal of the Academy of Marketing Science, 32*(1), 3–19.

Maignan, I., Ferrel, O., & Ferrel, L. (2005). A Stakeholder Model for Implementing Social Responsibility in Marketing. *European Journal of Marketing, 29*(9/10), 956–977.

Maignan, I., Ferrell, O. C., & Hult, G. T. M. (1999). Corporate citizenship: Cultural antecedents and business benefits. *Journal of the Academy of Marketing Science, 27*(4), 455–469.

Maloni, M. J., & Brown, M. E. (2006). Corporate social responsibility in the supply chain: An application in the food industry. *Journal of Business Ethics, 68*, 35–52.

Maon, F., Swaen, V., & Lindgreen, A. (2006). Impact of CSR commitments and CSR communication on diverse stakeholders: the case of IKEA. In C. B. Bhattacharya, D. Levine, & N. C. Smith (Eds.), *Corporate responsibility and global business: Implications for corporate and marketing strategy, Conference proceedings*. London Business School.

Margolis, J., & J. Walsh: (2003). Misery Loves Companies: Rethinking Social Initiatives by Business. *Administrative Science Quarterly, 48*, 268–305.

Margolis, J. D., & Walsh, J. P. (2001). Social Enterprise Series No. 19—Misery Loves Companies: Whither Social Initiatives by Business? Harvard Business School Working Paper Series, No. 01-058.

Markides, C. (2008). *Game-changing Strategies: How to create new market space in established industries by breaking the rules*, Jossey-Bass.

Marrewijk, M. van (2003). Concepts and definitions of CSR and corporate sustainability: Between agency and communion. *Journal of Business Ethics, 44*(2–3), 95–105.

Martinuzzi, A., Gisch-Boie, S., & Wiman, A. (2010). Does corporate responsibility pay off? Exploring the links between CSR and competitiveness in Europe's industrial sectors, final report to the European Commission, Directorate-General for Enterprise and Industry, Wien.

Matten, D., & Moon, J. (2008). Implicit and Explicit CSR: A Conceptual Framework for a Comparative Social Responsibility. *Academy of Management Review, 33*(2), 404–424.

McAdam, R., & Leonard, D. (2003). Corporate social responsibility in a total quality management context: Opportunities for sustainable growth, *Corporate Governance, 3*(4), 36–45.

McCallum, S., Schmid, M. A., & Price, L. (2013). CSR: A case of employee skills-based volunteering. *Social Responsibility Journal, 9*(3), 479–495.

McDonough, E. F. (2000). Investigation of factors contributing to the success of cross-functional teams. *Journal of Product Innovation Management, 17*, 221–235.

McKinsey Global Institute. (2011). Resource revolution: Meeting the world's energy, materials, food, and water needs. Report.

McWilliams, A., & Siegel, D. S. (2000). Corporate social responsibility and financial performance: correlations or misspecifications? *Strategic Management Journal, 21*, 603–609.

McWilliams, A., & Siegel, D. S. (2011). Creating and capturing value: Strategic corporate social responsibility, resource-based theory, and sustainable competitive advantage. *Journal of Management, 37*(5), 1480–1495.

Melcrum. (2006). *Engaging Employees in Corporate Responsibility: How the World's Leading Companies Embed CR in Employee Decision-making*. London, UK: Author.

Melé, D. (2002). Not only stakeholders interests: the firm oriented towards the Common Good, In S. A. Cortright & M. J. Naughton, (Eds.), *Rethinking the purpose of business, interdisciplinary essays from catholic social tradition* (pp. 190–214). Notre Dame University Press, Notre Dame.

Melo, T., & Garrido-Morgado, A. (2012). Corporate reputation: A combination of social responsibility and industry. *Corporate Social Responsibility and Environmental Management, 19*, 11–31.

Melynyte, O., & Ruzevicius, J. (2008). The study of interconnections of corporate social responsibility and human resource management. *Economics and management, 13*, 817–823.

Meyskens, M., Carsrud, A. L., & Cardozo, R. (2010). The impact of resources on the success of social entrepreneurship organizations: The symbiosis of entities in the social engagement network. *Entrepreneurship & Regional Development, 22*(5), 425–455.

Miles, M., Munilla, L. S., & Darroch, J. (2006). The Role of Strategic Conversations with Stakeholders in the Formation of Corporate Social Responsibility Strategy. *Journal of Business Ethics, 69*(2), 195–205.

Miller, D. (1996). Configurations Revisited. *Strategic Management Journal, 17*, 505–512.

Miller, T. L., & Wesley, C. L. (2010). Assessing mission and resources for social change: An organizational identity perspective on social venture capitalists' decision criteria. *Entrepreneurship Theory and Practice, 34*(4), 705–733.

Mirvis, P. (2012). Employee Engagement and CSR: Transactional, relational, and developmental approaches. *California Management Review, 54*(4), 93–117.

Mitchell, R. K., Agle, B. R., & Wood, D. J. (1997). Toward a theory of stakeholder identification and salience: defining the principle of who and what really counts. *Academy of Management Review, 22*(4), 853–86.

Morimoto, R., Ash, J., & Hope, C. (2005). Corporate Social Responsibility Audit: From Theory to Practice. *Journal of Business Ethics, 62*, 315–325.

Morris, M., Schindehutte, M., & Allen, J. (2005). The entrepreneur's business model: toward a unified perspective. *Journal of Business Research, 58*, 726–735.

Morsing, M., & Vallentin, S. (2006). CSR and stakeholder involvement: The challenge of organisational integration. In: A. Kakabadse & M. Morsing (Eds.) *Corporate social responsibility: Reconciling aspiration with application* (pp. 245–254). Palgrave Macmillan.

Morsing, M., & Perrini, F. (2009). CSR in SMEs: do SMEs matter for the CSR agenda? *Business Ethics: A European Review, 18*(1), 1–6.

Morsing, M. (2006). Drivers of Corporate Social Responsibility in SMEs. Conference paper. Presented at International CSR conference at Copenhagen Business School.

Morsing, M., & Schultz, M. (2006). Corporate social responsibility communication: stakeholder information, response and involvement strategies. *Business Ethics: A European Review*, *15*(4), 323–338.

Morsing, M., & Thyssen, C. (2003). *Corporate values and responsibility—the case of Denmark.* Samfundslitteratur.

Morsing, M., Nielsen, K. U., & Schultz, M. (2004). Komm ikationsstrategi for Social ansvarlighed. Reputation Institute.

Morsing, M., Schultz, M., & Nielsen, K. U. (2008). The "Catch 22" of communicating CSR: 313 findings from a Danish study. *Journal of Marketing Communications*, *14*(2), 97–111.

Muller, A. (2006, Apr–Jun). Global Versus Local CSR Strategies. *European Management Journal*, *24*(2/3), 189–198.

Murillo, D., & Lozano, J. (2006). SMEs and CSR: An approach to CSR in their own words. *Journal of Business Ethics*, *67*, 227–240.

Nair, M. (2011). Understanding and measuring the value of social media. *The Journal of Corporate Accounting & Finance*, *22*(3), 45–51.

Neergaard, P., Crone Jensen, E., & Thusgaard Pedersen, J. (2009). Partnerskaber mellem virksomheder ogfrivillige organisationer: En analyse af omfang, typer, muligheder og faldgrupper i partnerskaber. Erhvervs- og Selskabsstyrelsen.

Nelson, D. (1980). *Frederick W. Taylor and the Rise of Scientific Management.* Madison: University of Wisconsin Press.

Newman, L. (2005). Uncertainty, innovation, and dynamic sustainable development. *Sustainability, Science, Practice & Policy*, *1*(2), 25–31.

Nicholls, A. (2010). The legitimacy of social entrepreneurship: Reflexive isomorphism in a pre-paradigmatic field. *Entrepreneurship Theory and Practice*, *34*(4), 611–633.

Nidumolu, R., Prahalad, C. K., & Rangaswami, M. (2009). Why sustainability is now the key driver of innovation. *Harvard Business Review*, *87*(9), 56–64.

Nielsen, A. E., & Thomsen, C. (2009). Investigating CSR communication in SMEs: a case study among Danish middle managers. *Business Ethics: A European Review*, *18*(1), 83–93.

Norrgren, F., & Schaller, J. (1999). Leadership style: its impact on cross-functional product development. *Journal of Product Innovation Management*, *16*, 377–384.

Nygaard, H. H., & Aabling, C. (2011). Ansvarlig leverandørstyring—en saga blot? Published July 26th, 2011 at www.csr.dk/leverandor

O' Dwyer, B. (2003). Conceptions of corporate social responsibility: The nature of managerial capture. *Accounting, Auditing and Accountability Journal, 16*(4), 523–557.

O'Connell, B. (1999). *Civil society: The underpinnings of American democracy.* Hanover: University Press of New England.

O'Riordan, L., & Fairbrass, J. (2008). Corporate social responsibility (CSR): models and theories in stakeholder dialogue. *Journal of Business Ethics, 83*(4), 745–758.

O'Rourke, D. (2003). Outsourcing Regulation: Analyzing Nongovernmental Systems of Labor Standards and Monitoring. *Policy Studies Journal, 31*(1), 1–29.

OECD. (2004). OECD principles of corporate governance. Report. OECD Publications Service.

OECD. (2013). Anti-Corruption Ethics and Compliance Handbook for Business. Report. OECD.

Orlitzky, M., Schmidt, F. L., & Rynes, S. L. (2003). Corporate social and financial performance: A meta-analysis. *Organization Studies, 24*, 403–441.

Osterwalder, A. (2004). The Business Model Ontology. In: A Proposition in a Design Science Approach. Universiteì de Lausanne, Lausanne.

Ostrom, E., Schroeder, L., & Wynne, S. (1993). *Institutional incentives and sustainable development: Infrastructure policies in perspective*. Boulder: Westview Press.

Otto Scharmer, C. (2007). *Theory U: Leading from the Future as it Emerges.* The Society for Organizational Learning. Cambridge, USA.

Panapanaan, V. (2006). *Exploration of the Social Dimension of Corporate Responsibility in a Welfare State.* Lappeenranta University of Technology.

Parker, C. (2002). *The Open Corporation.* Cambridge University Press, Cambridge.

Pavelin, S., & Porter, L. A. (2008). The Corporate Social Performance Content of Innovation in the U.K. *Journal of Business Ethics, 80*,(1) 711–725.

Pedersen, E. R., & Neergaard, P. (2007). The bottom line of CSR: A different view. In F. D. Hond, F. G. A. D. Bakker, & P. Neergaard (Ed.), *Managing corporate social responsibility in action: talking, doing and measuring* (77–91). Aldershot: Ashgate Publishing, Ltd.

Pedersen, G. R. E., Pedersen, T. J., & Jacobsen, Ø. P. (2011). Partnerskaber mellem virksomheder og NGOer, *Ledelse og Erhvervsøkonomi*, Vol. 4.

Pennings, J. (1987). Structural Contingency Theory: A Multivariate Test. *Organization Studies, 8*, 223–240.

Perkman, M., & Spicer, A. (2010). What are business models? In: N. Phillips, G. Sewell, & D. Griffiths, (Eds.), *Research in the sociology of organizations*. Bingley, United Kingdom: Emerald Group Publishing Ltd.

Perrini, F. (2006). SMEs and CSR Theory: Evidence and Implications from an Italian Perspective. *Journal of Business Ethics, 67*(3), 305–316.

Perrini, F., Russo, A., & Tencati, A. (2007). CSR Strategies of SMEs and Large Firms. Evidence from Italy. *Journal of Business Ethics, 74*, 285–300.

Pfeffer, J. (2010). Building sustainable organizations: The human factor. *Academy of Management Perspectives, 24*(1), 34–45.

Pfeffer, J., & Salancik, G. R. (1978). *The External Control of Organizations*, Stanford, CA: Stanford University Press.

Philips, R., Freeman, E. R., & Wicks, A. C. (2003). What stakeholder theory is not, *Business Ethics Quarterly, 13*(4), 479–502.

Pinkston, T. S., & Carroll, A. B. (1994). Corporate Citizenship Perspectives and Foreign Direct Investment in the U.S. *Journal of Business Ethics, 13*(3), 157–169.

Pless, N., Maak, T., & Stahl, G. K. (2011). Developing responsible global leaders through international service-learning programs: The Ulysses experience. *Academy of Management Learning and Education, 10*(2), 237–260.

Poncelet, E. C., (2001). A kiss here and a kiss there: Conflict and collaboration in environmental partnerships, *Environmental Management, 27*(1), 13–25.

Porter, M. E. and Kramer, M. R. (2002). The competitive advantage of corporate philanthropy, Harvard Business Review, Vol. 80, No 12, pp. 56–69.

Porter, M. E., & Kramer, M. R. (2006). Strategy and society: The link between competitive advantage and corporate social responsibility. *Harvard Business Review, 84*(12) (December 2006).

Prasad, M., Kimeldorf, H., Meyer, R., & Robinson, I. (2004). Consumers of the World Unite: A market-based response to sweatshops. *Labor Studies Journal, 29*(3), 57–79.

PWC, 14th Annual Global CEO Survey 2011—In-depth story. Growth reimagined Prospects in emerging markets drive CEO confidence.

Quist, J., & Vergragt, P. (2006). Past and future of backcasting: The shift to stakeholder participation and a proposal for a methodological framework. *Futures, 38*(9), 1027–1045.

Ranganathan, J. (1998). Sustainability rulers: Measuring corporate environmental and social performance. Sustainable Enterprise Perspectives. World Resource Institute.

Rass, M., Dumbach, M., Danzinger, F., Bullinger, A. C., & Moeslien, K. M. (2013). Open Innovation and Firm Performance: The Mediating Role of Social Capital. *Creativity and Innovation Management, 22*(2), 177–194.

Ritter, T., & Andersen, H. (2014). A relationship strategy perspective on relationship portfolios: Linking customer profitability, commitment, and growth potential to relationship strategy. *Industrial Marketing Management*, *43*(6), 1005–1011.

Robinson, J. (2006). Navigating social and institutional barriers to markets: How social entrepreneurs identify and evaluate opportunities. In J. Mair, J. Robinson & K. Hockerts (Eds.), *Social Entrepreneurship* (95–120). New York, NY: PalgraveMacmillan.

Roepstorff, A., & Serpa, L. B. (2005). "Katalog over CSR værktøjer". Erhvervs- og Selskabsstyrelsen.

Rosthorn, J. (2000). Business ethics auditing—More than a stakeholder's toy. *Journal of Business Ethics,* 27, 9–19.

Rother, M., & Shook, J. (1999). *Learning to See: Value-stream mapping to create value and eliminate muda*. Brookline, MA: Lean Enterprise Institute. ISBN 0-9667843-0-8.

RQ. (2004). Reputation Institute. www.reputationinstitute.com

Ruebottom, T. (2011). Counting social change: Outcome measures for social enterprise. *Social Enterprise Journal, 7*(2), 173–182.

Russo, A., & Tencati, A. (2009). Formal vs. Informal CSR Strategies: Evidence from Italian Micro, Small, Medium-sized, and Large Firms. *Journal of Business Ethics*, *85*(Supplement 2), 339–353.

Rydberg, T. (1995). Cleaner products in the Nordic countries based on the life cycle assessment approach: The Swedish product ecology project and the Nordic project for sustainable product Development. *Journal of Cleaner Production*, *3*(1), 101–105.

Sachs, S. M., Mauer, M., Rühli, E., & Hoffmann, R. (2006). Corporate social responsibility from a 'stakeholder view' perspective: CSR implementation by a Swiss mobile telecommunication provider, *Corporate Governance*, *6*(4), 506–515.

Salamon, L. M., Sokolowski, S. W. & List, R. (2003). Report: Global Civil Society—An Overview. The Johns Hopkins Comparative Nonprofit Sector Project. The John Hopkins University, Instiute for Policy Studies & Center for Civil Society Studies.

Sandberg, K. D. (2002). Is It Time to Trade In Your Business Model? Harvard Management Update, 7, 3.

Sarkis, J. (1995). Manufacturing strategy and environmental consciousness. *Technovation*, *15*(2), 79–97.

Sarkis, J. (1998). Evaluating environmentally conscious business practices. *European Journal of Operational Research*, *107*(1), 159–174.

Sarkis, J. (2001). Manufacturing's role incorporate environmental sustainability—Concerns for the new millennium. *International Journal of Operations & Production Management*, *21*(5/6), 666–686.

Schaltegger, S., & Wagner, M. (2011). Sustainable entrepreneurship and sustainability innovation. Categories and interactions. *Business Strategy and the Environment*, *20*(4), 222–237.

Schaltegger, S., Lüdeke-Freund, F., Hansen, E. G. (2012). Business cases for sustainability—the role of business model innovation for corporate sustainability. International Journal of Innovation and Sustainable Development 6 (2), 95–119.

Schmiemann, M. (2008). Enterprises by size class—overview of SMEs in the EU. Industry, Trade and Services, Eurostat, Statistics in focus, nr. 31.

Schwartz, B., & Tilling, K. (2009). ISO-lating corporate social responsibility in the organizational context: A dissenting interpretation of ISO 26000. *Corporate Social Responsibility and Environmental Management*, *16*(5), 289–299.

Schwartz, M. S. (2002). A code of ethics for corporate code of ethics. *Journal of Business Ethics, 41*, 27–43.

Scott, W. R. (1987). *Organizations: Rational, Natural, and Open Systems. Englewood.* Cliff, NJ: Prentice-Hall.

Seuring, S, & Müller, M. (2008). Core issues in sustainable supply chain management: a Delphi study. *Business Strategy and the Environment, 17*(8), 455–466.

Sheppeck, M. A., & Militello, J. (2000). Strategic HR configurations and organizational performance. *Human Resource Management*, *39*(1), 5–16.

Short, J. C., Moss, T. W., & Lumpkin, G. T. (2009). Research in social entrepreneurship: Past contributions and future opportunities. *Strategic Entrepreneurship Journal*, *3*(2), 161–194.

Siggelkow, N. (2001). Change in the Presence of Fit: The Rise, the Fall, and the Renaissance of Liz Claiborne. *Academy of Management Journal, 44*, 838–857.

Silberhorn, D., & Warren, R. C. (2007). Defining corporate social responsibility: A view from big companies in Germany and the UK. *European Business Review, 19*(5), 352–372.

Siltaoja, M. E. (2013). Revising the corporate social performance model—towards knowledge creation for sustainable development. *Business Strategy and the Environment, 23*(5), 289–302.

Simmons, J. (2009). Both sides now: Aligning external and internal branding for a socially responsible era. *Marketing Intelligence & Planning, 27*(5), 681–697.

Sones, M., & Grantham, S. (2009). Communicating CSR via Pharmaceutical Company Websites. Evaluating Message Frameworks for External and Internal Stakeholders. *Corporate Communications: an International Journal, 14*(2), 144–157.

Spence, L., & Schmidpeter, R. (2002). SMEs, social capital and the common good, *Journal of Business Ethics, 45*, 93–108.

Spencer, L. J., & Rutherfoord, R. (2003). Small business and empirical perspectives in business ethics: Editorial. *Journal of Business Ethics, 47*(1), 1–5.

Sroufe, R., Liebowitz, J., & Sivasubramaniam, N. (2010). Are you a leader or a laggard? HR's role in creating a sustainability culture. *People & Strategy, 33*(1), 35–42.

Stanwick, P. A., & Stanwick, S. D. (1998). The relationship between corporate social performance, and organizational size, financial performance, and environmental performance: an empirical examination. *Journal of Business Ethics, 17*(2), 195–204.

Steketee, D. M. (2010). Disruption or sustenance? An institutional analysis of sustainable business network in west Michigan, In J. Sarkis, D. V. Brust, & J. J. Cordeiro, (Eds). *Facilitating sustainable innovation through collaboration: A multi-stakeholder perspective*. Springer.

Steurer, R. (2010). The role of governments in corporate social responsibility: characterising public policies on CSR in Europe. *Policy Sciences, 43*(1), 49–72, DOI: 10.1007/s11077-009-9084-4.

Stubbs, W., & Cocklin, C. (2008). Conceptualizing a sustainability business model, *Organization & Environment, 21*(2), 103–127.

Sun, J., Han, B., Ekwaro-Osire, S., & Zhang, H.-C. (2003). Design for environment: methodologies, tools, and implementation. *Journal of Integrated Design and Process Science, 7*(1), 59–75.

Swanson, D. L. (1995). Addressing a Theoretical Problem by Reorienting the Corporate Social Performance Model, *Academy of Management Review, 20*(1), 43–64.

Tapping, D., & Shuker, T. (2007). Lean i service & administration. DI september 2007. Drejebog med 8 trin til kortlægning og forbedring af værdistrømme.

Tapscott, D. (2008). *Grown up digital—how the Net generation is changing your world.* McGraw Hill.

Taran, Y., Boer, H., & Lindgren, P. (2013). A Business Model Innovation Typology. *Journal of Decision Science.*

Taran. Y. (2011). Re-thinking it All: Overcoming Obstacles to Business Model Innovation Center for Industrial Production. Ph.D. Thesis, Aalborg University.

Tata, J., & Prasad, S. (2015). CSR Communication: An Impression Management Perspective. *Journal of Business Ethics, 132*, 765–778.

Taylor, F. W. (1911). *The Principles of Scientific Management.* Harper & Brothers.

Teece, D. J. (2010). Business models, business strategy and innovation. *Long Range Planning*, *45*(2–3), 172–194.

Thompson, J. K., & Smith, H. L. (1991). OSocial Responsibility and Small Business: Suggestions for Research. *Journal of Small Business Management, 29*(January), 30–44.

Tilley F: 2000, Small Firm Environmental Ethics: How Deep Do They Go? *Business Ethics: A European Review*, *9*(1), 31–41.

TNS Gallup. (2005). Mapping of CSR-activities among small and medium-sized companies.

Tracey, P., & Jarvis, O. (2007). Toward a theory of social venture franchising. *Entrepreneurship Theory and Practice*, *31*(5), 667–686.

Tracey, P., & Phillips, N. (2007). The distinctive challenge of educating social entrepreneurs: A postscript and rejoinder to the special issue on entrepreneurship education. *Academy of Management Learning & Education*, *6*(2), 264–271.

Tukker, A., & Tischner, U. (Eds.) (2006). *New Business for Old Europe. Product-service Development, Competitiveness and Sustainability.* Greenleaf, Sheffield.

Turban, D. B., & Greening, D. W. (1997). Corporate Social Performance and Organizational Attractiveness to Prospective Employees. *Academy of Management Journal*, *40*(3), 658–672.

Uhl-Bien, M., Marion, R., & McKelvey, B. (2007). Complexity Leadership Theory: Shifting Leadership from the Industrial Age to the Knowledge Era. *The Leadership Quarterly 18*(4), 298–318.

Ulrich, D. (1998). A new mandate for human resources. *Harvard Business Review*, January–February, 124–139.

Urip, S. (2010). *CSR Strategies—corporate social responsibility—for a competitive edge in emerging markets.* John Wiley & Sons.

Uusi-Rauva, C., & Nurkka, J. (2010). Effective internal environment-related communication—An employee perspective. *Corporate Communications: An International Journal, 15*(3), 299–314.

Vaaland, T. I., & Heide, M. (2008). Managing corporate social responsibility: lessons from the oil industry. *Corporate Communications: An International Journal*, *13*(2), 212–225.

Vachon, S., & Klassen, R. D. (2008). Environmental management and manufacturing performance: the role of collaboration in the supply chain. *International Journal of Production Economics, 111*(2), 299–315.

Valle, R., Martin, F., Romero, P. M., & Dolan, S. L. (2000). Business strategy, work processes and human resource training: are they congruent? *Journal of Organisational Behavior*, *21*, 283–297.

Vallentin, S. (2001). *Pensionsinvesteringer, etik og offentlighed—en systemteoretisk analyse af offentlig meningsdannelse*. Samfundslitteratur.

Vallentin, S. (2011). *Afkastet og anstændigheden—social ansvarlighed i kritisk belysning,* Samfundslitteratur, Frederiksberg, Denmark.

Van de Vrande, V., de Jong, J. P. J., Vanhaverbeke, W., & de Rochemont, M. (2009). Open Innovation in SMEs: Trends, Motives and Management Challenges. *Technovation*, 29, 423–37.

Van Tulder, R., & van der Zwart, A. (2006). *International Business-Society management: Linking corporate responsibility and globalization*. Routledge: Abongdon, Oxon.

Veleva, V., & Ellen Becker, M. (2001). Indicators of sustainable production: framework and methodology. *Journal of Cleaner Production* 9, 519–549.

Ven Bert van de & Jeurissen, R. (2005). Competing responsibly. *Business Ethics Quarterly, 15*(2), 299–317.

Venkatraman, N., & Camillus, J. C. (1984). Exploring the Concept of ‘Fit’ in Strategic Management. *Academy of Management Review, 9*(3), 513–525.

Verhaugen, G. (2003). Opportunity and responsibility. How to help more small businesses to integrate social and environmental issues into what they do, European Commission, Directorate General for Enterprise and Industry.

Verstraete, T., & Jouison-Lafitte, E. (2011). A conventionalist theory of the business model in the context of business creation for understanding organizational impetus. *Management International/International Management/Gestiôìn International, 15*, 109–124.

Vogel, D. J. (2005). Is there a market for virtue? The business case for corporate social responsibility. *California Management Review, 47*(4), 19–45.

Vogel, D. J. (2005). *The market for virtue: The potential and limits of corporate social responsibility*. Brookings Institution Press.

Volery, T., & Hackl, V. (2010). The promise of social franchising as a model to achieve social goals. In A. Fayolle & H. Matlay (Eds.), *Handbook of Research on Social Entrepreneurship* (155–179). Cheltenham: Edward Elgar.

Von Hippel, E. (2005). *Democratizing Innovation.* The MIT Press.

Waddock, S. A., & Graves, S. B. (1997). The corporate social performance—financial social link, *Strategic Management Journal, 18*(4), 303–319.

Wagner, M. (2010). Corporate Social Performance and Innovation with High Social Benefits: A Quantitative Analysis. *Journal of Business Ethics, 94*, 581–594.

Waldman, D. A., & Siegel, D. (2008). Defining the Socially Responsible Leader. *The Leadership Quarterly, 19*(1), 117–131.

Waldman, D. A., & Galvin, B. M. (2008). Alternative perspectives of responsible leadership. *Organizational Dynamics, 37*(4), 327–341.

Walker, H., Di Sisto, L., & McBain, D. (2008). Drivers and barriers to environmental supply chain management practices: Lessons from the public and private sectors. *Journal of Purchasing and Supply Management, 14*(1), 69–85.

Walumbwa, F. O., Avolio, B. J., Gardner, W. J., Wernsing, T. S., & Peterson, S. J. (2008). Authentic Leadership: Development and Validation of a Theory-Based Measure. *Journal of Management, 34*(1), 89–126.

Waters, R., Burnett, E., Lammb, A., & Lucas, J. (2009). Engaging stakeholders through social networking: how nonprofit organizations are using Facebook. *Public Relations Review, 35*(2), 102–106.

WBCSD (2000). Corporate Social Responsibility: Making good business sense. World Business Council for Sustainable Development. ISBN 2-94-024007-8.

WCED World Commission on Environment and Development. (1987). *Our Common Future.* Oxford University Press.

Weber, M. (2008). The business case for corporate social responsibility: A company-level measurement approach for CSR. *European Management Journal, 26*(4), 247–261.

Welch, M., & Jackson, P. R. (2007). Rethinking internal communication: a stakeholder approach. *Corporate Communications: An International Journal, 12*(2), 177–98.

Welford, R. (2007). Corporate Social Responsibility in Europe and Asia: Critical Elements and Best Practice. *Journal of Corporate Citizenship*, Spring.

Welford, R. (2004). Corporate Social Responsibility in Europe and Asia: Critical Elements and Best Practice, *Journal of Corporate Citizenship, 13*, 31–47.

Wenlong, Y., Bao, Y., & Verbeke, A. (2011). Integrating CSR Initiatives in Business: An Organizing Framework. *Journal of Business Ethics, 101*(1), 75–92.

Werre, M. (2003). Implementing Corporate Responsibility—The Chiquita Case. *Journal of Business Ethics, 44*(2–3), 247–260.

Werther, W., & Chandler, D. (2005). Strategic corporate social responsibility as global brand insurance. *Business Horizons, 48*, 317–324.

West, J., & Bogers, M. (2014) Leveraging External Sources of Innovation: A Review of Research on Open Innovation. *Journal of Product Innovation Management, 31*, 814–31.

Westley, F., & Vredenburg, H. (1996). Sustainability and the Corporation: Criteria for Aligning Economic Practice with Environmental Protection. *Journal of Management Inquiry, 5*(2), 104–119.

Weybrecht, G. (2010). *The sustainable MBA: The manager's guide to green business.* Wiley.

Wheeler, M. A., & Thomas, D. (1997). 3M: Negotiating Air Pollution Credits (A). *Harvard Business School Case* 897–134.

White, A. (2005). Fade, integrate or transform? The future of CSR. *Business for Social Responsibility*, Issue Paper. www.jussemper.org

Willer, H., & Kilcher, L. (Eds.) (2011). The World of Organic Agriculture. Statistics and Emerging Trends 2011. IFOAM, Bonn, & FiBL, Frick.

Williamson, D., Wood, G. L., & Ramsay, J. (2006). Drivers of Environmental Behaviour in Manufacturing SMEs and the Implications for CSR. *Journal of Business Ethics, 67*(3), 317–330

Williamson, O. E. (1995). Chester Barnard and the Incipient Science of Organization. In O. E. Williamson (Ed.) *Organization theory: from chester barnard to the present and beyond,* (pp. 172–206). Oxford University Press.

Winston, A. (2014). GE Is Avoiding Hard Choices About Ecomagination. August 1, 2014. *Harvard Business Review*. https://hbr.org/2014/08/ges-failure-of-ecomagination#

Wittenberg, J., Harmon, J., Russel, W. G., & Fairfield, K. D. (2007). HR's role in building a sustainable enterprise: Insights from some world's best companies. *Human Resource Planning, 30*(1), 10–20.

Wright, P., & Ferris, S. (2000). Agency conflict and corporate strategy: The effects of divestment on corporate value. *Strategic Management Journal, 18*, 77–83.

Wright, P. M., & McMahan, G. C. (1992). Theoretical perspectives for strategic human resource management. *Journal of Management, 18*, 295–320.

Wright, P. M., & Boswell, W. R. (2002). Desegregating HRM: a review and synthesis of micro and macro human resource management research. *Journal of Management, 28*, 247–276.

Yaziji, M. (2004). Turning gadflies into allies, *Harvard Business Review, 82*(2), 110–115.

Yaziji, M., & Doh, J. (2009). *NGOs and Corporations: Conflict and Collaboration.* Cambridge: Cambridge University Press.

Yu, X. (2008). Impacts of Corporate Code of Conduct on Labor Standards: A Case Study of Reebok's Athletic Footwear Supplier Factory in China. *Journal of Business Ethics, 81*(3), 513–529.

Yunus, M., Moingeon, B., Lehmann-Ortega, L. (2010). Building social business models: lessons from the Grameen experience. *Long Range Planning, 43*, 308–325.

Zahra, S., Gedajlovic, E., Neubaum, D., & Shulman, J. (2009). A typology of social entrepreneurs: Motives, search processes and ethical challenges. *Journal of Business Venturing, 24*(5), 519–532.

Zink, K. J., Steimle, U., & Fisher, K. (2008). Human factors, business excellence and corporate sustainability: Differing perspectives, joint objectives. In K. J. Zink, (Ed.). *Corporate sustainability as a challenge for comprehensive management*. Physica-Verlag.

Zott, C. & Amit, R. (2007). Business model design and the performance of entrepreneurial firms. Organization Science 18, 181–199.

Zott, C., Amit, R., & Massa, L. (2011). The business model: Recent developments and future research. *Journal of Management, 37*(4), 1019–1042.

Index

About the Author

Annabeth Aagaard (Ph.D., M.Sc.) has over twenty years of experience working as a manager, management specialist, and academic researcher across strategy, management, CSR/sustainability, business modeling, and innovation among numerous large and global companies and universities.

She is an associate professor at the Department of Business Development and Technology at Aarhus University in Denmark. She is the co-manager of the MBIT (Multi-Business-Model-Innovation & Technology) laboratory and of the Stanford Peace Innovation Lab. She holds a Ph.D. in Front End Innovation in Pharma and Biotech and has interests within the scientific fields of sustainability, innovation, business modeling, and strategic management drawing on her theoretical knowledge and practical experiences in the academic, public, and private sectors.

Through her elaborate network, she performs and provides empirical and applied research with the main objective to bridge the gap between theory and practice. She has authored and co-authored nine academic textbooks and management handbooks and several scientific journal articles on different innovation, business modeling, and CSR topics. Her research is published in many of the top-ranked scientific journals in management, sustainability and innovation. She is also an acknowledged public speaker and has published numerous public articles on key sustainability and innovation topics.

For Product Safety Concerns and Information please contact our EU
representative GPSR@taylorandfrancis.com
Taylor & Francis Verlag GmbH, Kaufingerstraße 24, 80331 München, Germany

www.ingramcontent.com/pod-product-compliance
Ingram Content Group UK Ltd.
Pitfield, Milton Keynes, MK11 3LW, UK
UKHW022210190926
475702UK00003B/293

* 9 7 8 8 7 9 3 3 7 9 7 9 4 *